ENGINEERING PROBABILISTIC DESIGN AND MAINTENANCE FOR FLOOD PROTECTION

ENGINEERING PROBABILISTIC DESIGN AND MAINTENANCE FOR FLOOD PROTECTION

EDITED BY

Roger COOKE
Delft University of Technology
Delft, The Netherlands

Max MENDEL
University of California
Berkeley, California, USA

and

Han VRIJLING
Delft University of Technology
Delft, The Netherlands

KLUWER ACADEMIC PUBLISHERS
DORDRECHT / BOSTON / LONDON

A C.I.P. Catalogue record for this book is available from the Library of Congress

ISBN-13: 978-1-4613-3399-9 e-ISBN-13: 978-1-4613-3397-5
DOI: 10.1007/978-1-4613-3397-5

Published by Kluwer Academic Publishers,
P.O. Box 17, 3300 AA Dordrecht, The Netherlands.

Kluwer Academic Publishers incorporates
the publishing programmes of
D. Reidel, Martinus Nijhoff, Dr W. Junk and MTP Press.

Sold and distributed in the U.S.A. and Canada
by Kluwer Academic Publishers,
101 Philip Drive, Norwell, MA 02061, U.S.A.

In all other countries, sold and distributed
by Kluwer Academic Publishers Group,
P.O. Box 322, 3300 AH Dordrecht, The Netherlands.

Printed on acid-free paper

CONTENTS

PREFACE

The First Conference on Engineering Probability in Flood Defense was organized by the Department of Mathematics and Informatics of the Delft University of Technology and the Department of Industrial Engineering and Operations Research of the University of California at Berkeley, and was held on June 1,2 1995 in Delft. Groups at Berkeley and Delft were both deeply engaged in modeling deterioration in civil structures, particularly flood defense structures. The plans for the conference were well under way when the dramatic floods in The Netherlands and California in the winter of 1994-1995 focused world attention on these problems.

The design of civil engineering structures and systems is essentially an example of decision making under uncertainty. Although the decision making part of the process is generally acknowledged, the uncertainty in variables and parameters in the design problem is less frequently recognized. In many practical design procedures the uncertainty is concealed behind sharp probabilistic design targets like 'once in a thousand years' combined with a standardized use of safety factors. The choice of these probabilistic design targets, however, is based on an assessment of the uncertainty of the variable under consideration, and on its assessed importance. The value of the safety factor is governed by similar considerations. Standard practice is simply accumulated experience and engineering judgment.

In light of the great number of civil engineering structures that function successfully, one may say that this standard practice has proven itself broadly satisfactory.

In probabilistic design the relation between the uncertainty of variables, the required safety level and the values of the safety coefficients is studied with explicit mathematical relations. While this is not necessary for the great bulk of civil engineering design problems, it is necessary for large unique projects involving considerable capital investment and large potential risk. For such problems there simply is no pre-existing body of experience and judgment. As these probabilistic design techniques become more streamlined and more familiar, we may expect the class of problems to which they may be profitably applied to grow. The studies in this book illustrate how probabilistic techniques have contributed to the solution of problems in the area of flood defense.

Solving complex problems (and flood defense is a complex problem) involves marshaling expertise from diverse disciplines. The major difficulty in doing this is overcoming barriers to communication. In reflecting on these problems, we opted for an experimental design for this conference. Main speakers were sought out who had recently done important applications-oriented work. For each main speaker's contribution, we sought three respondents representing different fields of expertise. Specifically we sought a mathematician, an engineer and a problem owner, and asked them to review the main contribution from their own specific field.

We were very pleased that the authors responded so enthusiastically to this challenge. The main contributions were available well in advance of the conference, and the reviewers prepared well. The reviews were incisive, and the discussions were brisk. The final versions of the main papers and reviews were written after the conference, but the cut and thrust of the discussions are preserved. The reader can judge for him-/her self the degree to which this novel approach succeeded. In any event, the participants agreed that intended communication did take place. Der Kiureghian and Li's final words to their reviewers capture the very constructive mood throughout this conference:

It has been a privilege to have these prominent engineers and researchers read the paper so carefully and prepare detailed comments that have indeed helped us improve our method and extend its scope of application. We have learned that a method that was originally developed for application to earthquake engineering problems is potentially useful in assessing the reliability of flood defense systems.

A few words about the contents of these proceedings may help orient the reader. A broad introductory chapter by M. Mendel shows the way to 'probabilize' traditional engineering approaches. The first main contribution (Van Noortwijk, Kok, Cooke) deals with sea-bed protection. The innovative modeling here is driven by two concerns: first, to capture the fundamental physical feature of monotonic deterioration, and second, to accommodate the model to available data. The use of Gamma processes in this regard provoked an excellent mathematical review of the literature by N. Singpurwalla, which in turn inspired J. van der Weide to find a new derivation of these processes.

The second main contribution (Chick, Shortle, Van Gelder, Mendel) uses river dynamics to predict extreme flood levels. The engineer and problem owner joined forces in this case to sketch the flood protection policies in the Netherlands as a backdrop for their review. The reader unfamiliar with the extensive Dutch experience will find this review most valuable.

The third main contribution (Li, Der Kiureghian) describes a new way of computing exceedance probabilities using the First Order Reliability Method (FORM). The reviewers in this case were able to deepen everyone's appreciation of this powerful method. This is reflected in the statement quoted above.

The final main contribution (Vrijling, Van Gelder) concerns the probabilistic design of berm breakwaters. They explicitly take account for uncertainty in dynamically stable structures in deriving optimal maintenance strategies. From their different perspectives, the reviewers all draw attention to the fact that real applications inevitably involve more assumptions than theoretical exercises.

Finally it is a pleasure to thank those who contributed to the realization of the conference and the preparation of the proceedings. Pieter van Gelder did the real work of keeping the authors on schedule and typesetting their contributions. He also managed the social agenda of the conference. Jan van Noortwijk helped with the organization in Delft. Steve Chick and John Shortle organized the contributions from the Berkeley side. The Department of Industrial Engineering and Operations Research at Berkeley provided support for editing the proceedings. Richard Jorissen gave us a guided tour of the Eastern Scheldt Storm Surge Barrier. Finally, financial support from the group of Stochastics, Statistics and Operations Research of the Department of Mathematics and Informatics at Delft, and the Road and Hydraulic Engineering Division and the Civil Engineering Division of the Dutch Ministery of Water Management (Rijkswaterstaat) is gratefully acknowledged.

Delft,
Berkeley,
July 2, 1996,

Roger Cooke,
Max Mendel,
Han Vrijling.

1

THE CASE FOR ENGINEERING PROBABILITY

Max Mendel

Department of Industrial Engineering and Operations Research, University of California, Berkeley, California 94720

ABSTRACT

Engineering probability concerns the derivation of probabilistic models from the physical laws, geometric constraints, and other engineering knowledge concerning an engineering system. This article shows how this can be done systematically. The engineering domain expertise is translated into a set of symmetries on the distribution, which is then translated into Bayesian likelihood models. One can then leverage off of Bayesian statistical theory to incorporate data into the model. The article uses a simple second-order system to demonstrate these techniques.

1 INTRODUCTION

This article argues for a separate discipline that could be called engineering probability. The emphasis in mainstream statistics is on data and how they are incorporated into a probabilistic model. To determine the initial model for an engineering system, however, extensive domain expertise is required. Although some of this expertise could be bypassed when there is an overabundance of data, this is often unavailable or prohibitively expensive in engineering applications. Design of flood defenses is a case in point. In the design phase, no data at all is available since nothing has been built yet and probability models for disaster have to be based on engineering knowledge. For maintenance of, say, dykes, one can rely on historical river levels. However, these are notoriously suspicious as land use can fundamentally change quite rapidly: for instance, urbanization increases run-off which increases river levels above historical levels. The solution here is to explicitly account for the effects of these changes

R. Cooke et al. (eds.), Engineering Probabilistic Design and Maintenance for Flood Protection, 1–22.

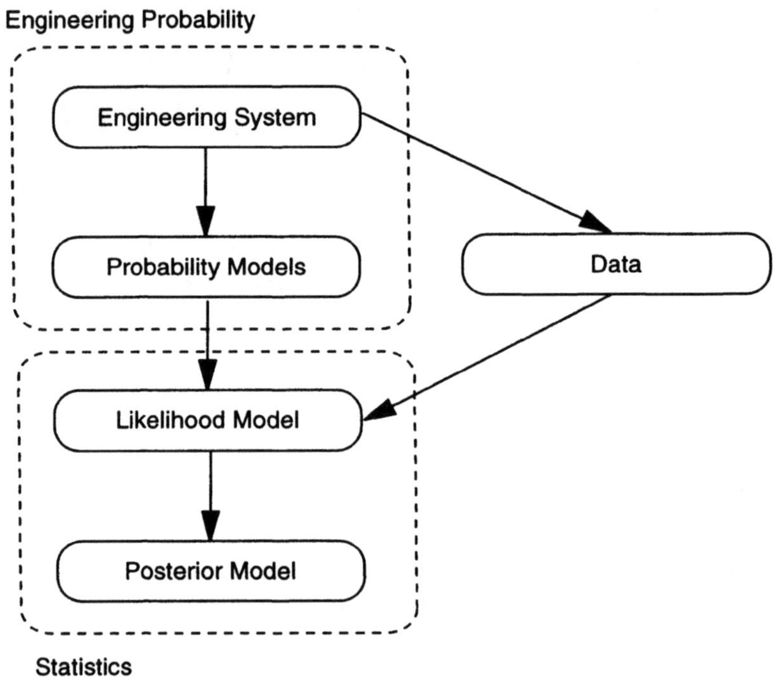

Figure 1 Complementarity of engineering probability and statistical theory.

in the probability models. Engineering probability would do this by deriving the probabilistic models directly from the physical structure of the underlying system, thereby eliminating the need for data.

Engineering probability, however, should complement existing statistical theory. In this way, both the engineering domain-expertise and any available data can be used to base decisions on. Figure 1 shows one way to do this. The figure breaks the process of determining a probability model up into four steps. Engineering probability refers to the first two: deriving a class of probability models that is consistent with the engineering domain-expertise. These can be translated into a statistical likelihood model. Statistics concerns the latter two: incorporating the data into the likelihood model.

In the next two Sections it is shown how the likelihood model can be derived from the domain expertise. We use a simple second order system. Its mechanical implementation is governed by Newton's three laws, the isotropicity

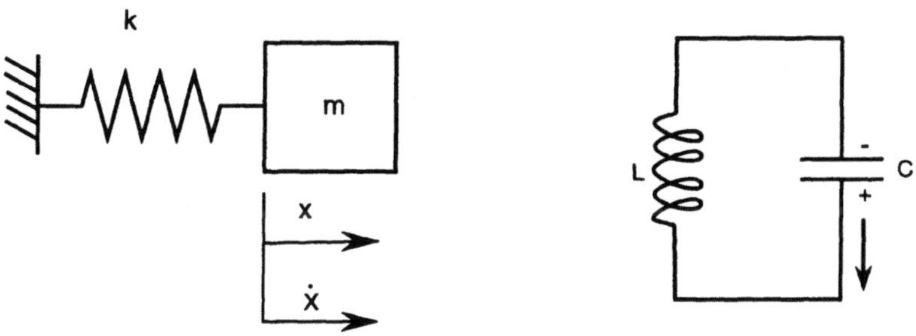

Figure 2 Undamped second order systems: left, a spring-mass system with spring constant k and mass m and, right, an LC circuit with inductance L and capacitance C. The position x corresponds to the charge on the capacitor and the velocity $\dot{x}$ to the current in the circuit.

of space, and the ideal spring law. Exploiting these laws, we derive a likelihood model that is related to the normal model. Although the complexity of this system is a far cry from those of the other articles in this volume, its simplicity allows us to carry the analysis through in depth, exactly and analytically.

2 SECOND-ORDER SYSTEM

Consider a second-order system without damping. Its mechanical and electrical implementations are shown in Figure 2. Here we focus on the mechanical implementation, the spring-mass system. It is a conservative system, meaning that the total energy:

$$U = \frac{1}{2}kx^2 + \frac{1}{2}m\dot{x}^2, \tag{1.1}$$

is constant, although it may be unknown. Here k is the spring constant, m the mass, x the position, and $\dot{x}$ the velocity.

Imagine the spring-mass system oscillating behind a wall so that its state cannot be directly observed. What is its position x at some random moment in time? This x is a random variable. Its distribution can be determined in two ways.

A data-based approach will sample the position at various random moments in time and base the distribution on these data. The results of such an analysis is shown in Figure 3. A second-order system was simulated on a computer and 1000 random samples of its position were taken. A bathtub-like shape appears

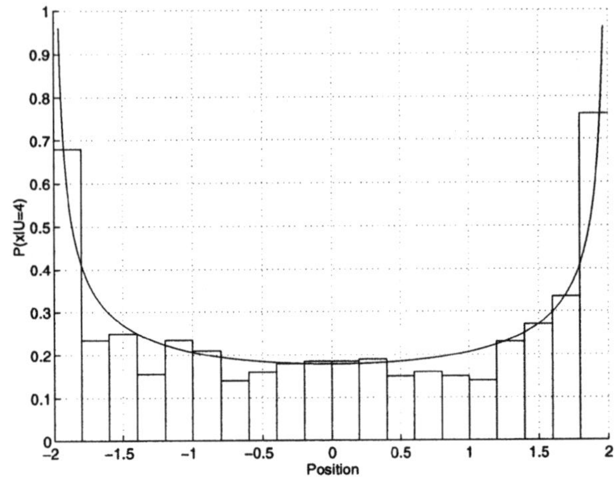

Figure 3 Histogram of the experimentally determined density of the position x of the spring-mass system. The continuous line is the engineering probability density.

for the density. It is apparently more likely to that the system is in an extreme position, then in its rest position in the middle $x = 0$.

However, it is possible to derive this distribution directly from Newton's law of motion ($F = m\ddot{x}$) and the ideal-spring law ($F = kx$), without taking a single sample. This is the engineering probability approach. Note first that the qualitative shape of the density in Figure 3 could have been predicted as follows: In the extreme position, the velocity changes direction and therefore has to vanish. In the middle at $x = 0$ the velocity is maximal. This means that the system spends more time in the extreme positions than in the middle. Therefore it is more likely to find it in the extreme position at some random moment in time. Quantitatively, it is true that the probability of finding the system at some random moment in time is inversely proportional to the absolute value of the velocity. From this is follows that the density for x is

$$p(x|U) \propto \frac{1}{\sqrt{U}} \left[1 - \frac{x^2}{2U/k}\right]^{-1/2}. \tag{1.2}$$

This density can be recognized as a finite-population version of the normal density. We will return to this point in a later section. It is graphed in Figure 4 for $U = 4$. In Figure 3 it is overlayed on the experimentally determined histogram.

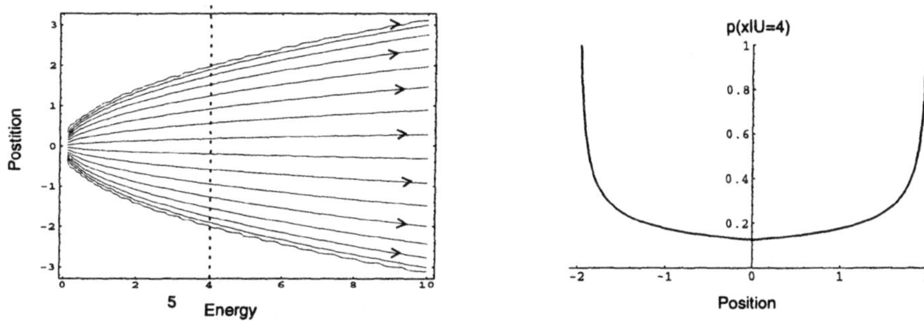

Figure 4 The conditional density 1-form of the position given the energy in (U,x) space (left) and the conditional density function with $U = 4$ (right). In the picture $k = 2$ for convenience.

The engineering-probability results have important advantages additional to obviating data collection. The expression in Equation (1.2) gives the density explicitly in terms of the unknown energy U and the spring constant k. With a data-based approach, a different set of data has to be collected for every value of the energy. This makes the latter much more of a descriptive model as opposed to the predictive nature of the engineering-probability model. This predictive feature is important in design. Imagine that the energy or the spring constant is some decision variable that can be set. It is then possible to predict the uncertainty in the position for various values of U or k and select an optimal value for them according to some design criterion.

Figure 4 shows the density in (1.2) as a function of U. To avoid a three-dimensional plot, we mark off intervals containing an equal amount of probability (1/16th in the figure). This gives a picture of conditional probability as a compressible fluid flow; as the flow lines diverge, the probability density becomes flatter. The density function for $U = 4$ is recovered by going along the dotted line and then graphing the amount of probability per unit of x. Differential-geometrically speaking, this picture gives the 1-form field in (U, x) coordinates that represents the conditional density. Shortle [5] contains a further discussion of these geometric methods.

The translation of the model into a Bayesian likelihood model is straightforward in this case. Think of the energy U as the statistical parameter. The conditional density $p(x|U)$ is then the likelihood function (albeit for only for a single data point x). $p(x|U)$ is known in the sense that it is determined by the physics of the system. The prior distribution, i.e., the distribution for U, is unknown in the sense that it is not determined by Newton law of motion and the ideal spring law. This is not surprising since these laws lead to a set of differential equation whose initial conditions which determine U are undetermined. Additional information concerning how the energy was imparted on this particular system is needed to determine the prior.

Notice that the emergence of a Bayesian model rather than a frequentist one is inevitable here. The energy parameter U is a function of the data x and $\dot{x}$. Insofar as these are random variables, U has to be a random variable and, hence, has to have a prior distribution. The situation for k is different. It is not a function of x or $\dot{x}$. Typically, k would be determined through some calibration experiments involving a set of known forces and displacements. k would be a function on such a calibration space and a random variable if the outcomes of the calibration experiment were unknown. The situation is similar for m. Here we will simply consider k and m to be decision variables.

The following pattern emerges from these considerations: In engineering one typically considers a class of systems; in this case all undamped second-order systems. These determine a class of probability models. This latter class can be represented by a Bayesian likelihood model. Its statistical parameter is in fact the conserved quantity or scalar invariant of the system. The likelihood summarizes the commonality among second-order systems and the prior identifies a particular system in motion. The analysis of such classes is useful as it gives us general tools rather than having to start from scratch for every particular system.

2.1 The role of symmetry

Much work has been done in statistics to derive likelihood models from symmetries imposed on distributions. This work started with de Finetti's famous derivation of the binomial model using exchangeability (see [2]) and reached a recent high in the work of Diaconis and Freedman (see, e.g., [4]). There is a interesting relation between this work and a more systematic approach for deriving engineering-probability models. This will be the subject of this subsection. It is somewhat technical, though.

Our approach is to translate the physical laws into geometric statements. These can then be used to derive the probability models. The tools and language of differential geometry are particularly useful here.

The starting point is to identify the basic probability space. This is the state space (or phase space) of the system. The state space is spanned by two variables: the position x and the velocity $\dot{x}$. This is $\Re^2$. Newton's laws give this space geometrical structure.

First, the existence of an inertial reference frame allow us to identify a unique position coordinate x, irrespective of the value of $\dot{x}$. Second, the fact that the body in rest persists in its motion allows us to identify the $\dot{x}$. This give the state space a fiber-bundle structure. For every position x there is a line (or fiber) of possible velocities at that position and vice versa.

Newton's laws also impose a vector field structure on this space. Time is defined to be such that motion looks simple. That is, we define t to be such that:

$$\dot{x} = \frac{dx}{dt}.$$

As time passes, the state $(x, \dot{x})$ moves along a path in the state space. The rate at which it does is describe by a vector with two components. The horizontal component is the change in the position per unit of time. Its magnitude is equal to $\dot{x} = \sqrt{(2U - kx^2)/m}$. The magnitude of the vertical component is the change in the velocity per unit of time. This is the acceleration of the mass and is given by Newton's second law

$$\ddot{x} = F/m = kx/m,$$

where in the last equality we used the spring law to determine F. At any point in state space, such a vector can be drawn. This is the vector field

$$\sqrt{\frac{2U - kx^2}{m}} \frac{\partial}{\partial x} + \frac{kx}{m} \frac{\partial}{\partial \dot{x}}.$$

This vector field is the structure we need to determine the probability model as will be shown presently.

Figure 5 shows the state space of the system with position x and velocity $\dot{x}$ coordinates. The physical laws are represented by the vector field. The vectors are tangent to the constant-energy ellipse: $U = m\dot{x}^2/2 + kx^2/2 = \text{constant}$.

We can interpret the vector field on the state space as the infinitesimal generator of a symmetry group acting on the space. This is the symmetry that can be used to determine the distributions and it is implied by the physics. If $k = m$ then this group is the group of rotations on $\Re^2$. This suggests looking at those distributions that are invariant under transformations of the state space under

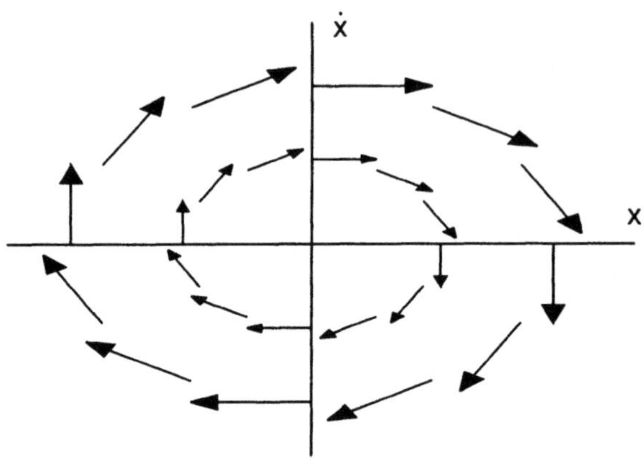

Figure 5 State space of a spring-mass system in x and $\dot{x}$ coordinates and the Hamiltonian vector field on the space.

the action of the vector field. These are, in a sense, the natural distributions for the system as they depend only on the physics underlying the system and are invariant under arbitrary coordinate transformations. In fact, this class of distributions is precisely the one given by the likelihood model presented above. An easy way to check this is to change coordinate to $y_1 = 2x/k$ and $y_2 = 2\dot{x}/m$ and then apply the known results for rotationally-symmetric distributions as given in [4].

Rather than invoking a known result, it is also possible to do the derivation directly or even purely geometrically. A random state for a given value of U is described by a differential 1-form field whose action on the vector field is constant. This is the differential-geometric statement of the invariance statement above. It follows immediately that the density for position is inversely proportional to the velocity and that of the velocity is inversely proportional to the acceleration. In Figure 6, these forms are indicated graphically by tick marks that enclose an infinitesimal unit amount of probability. The radial flow lines that connect the tick marks for various values of U represent the conditional probability 1-form in state space. This is an entirely coordinate-free picture of the distribution.

The differential-geometric statement relates to the age-old principle of indifference. This principle had become in disrepute on continuous space because, it was argued, that one could change a uniform density to any other through a change of variables. The problem from an engineering perspective is that the

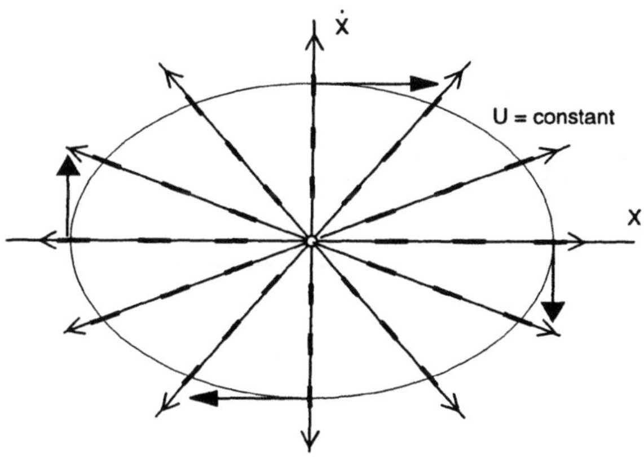

Figure 6 Radial flow lines representing the conditional distribution given U of the state and tick marks representing the distribution for a given value of U.

principle was applied to a particular coordinate, i.e., the vector field generated by that coordinate. This is indeed as arbitrary as the coordinate itself. By applying the principle with respect to a vector field generated by the physics, the principle becomes physically meaningful. Alternatively, the principle can be applied with respect to a coordinate that generates the vector field. This is also called a normal coordinate. In the example, this is time since

$$\frac{d}{dt} = \sqrt{\frac{2U - kx^2}{m}}\frac{\partial}{\partial x} + \frac{kx}{m}\frac{\partial}{\partial \dot{x}}.$$

This gives physical meaning to the statement that the distribution gives the system at a "random moment in time."

The derivation in this section suggests the following steps for deriving a model:

1. Identify a space in which every state of the system is a single point.
2. Identify the structure of this space that is determined by the physics and other engineering knowledge about the system.
3. Use the symmetries implied by the physical structure (and the principle of indifference or invariance) to determine the class of distributions on the space that is consistent with it.
4. Translate this class into a Bayesian parametric class.

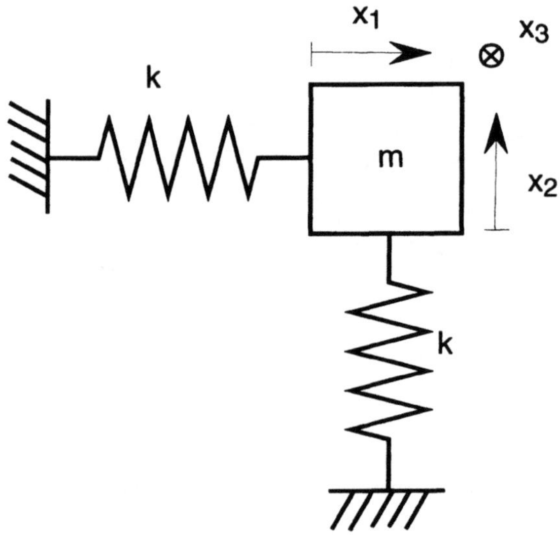

Figure 7 Three dimensional conservative second order system. The third spring is orthogonal to the springs in the plane and behind the mass.

In the spring-mass example, the space in step 1 is the state space of the system (a state of a dynamic system is defined so that, given the state, the entire future of the system is determined). The structure in step 2 is entirely determined by Newton's laws and the spring law. Time is defined through Newton's laws, so we could say that the system is in a random state with respect to these laws, rather than with respect to time. The symmetries in step 3 were rotational symmetries (up to a change in variables). On $\Re^\infty$ rotational symmetry yields the normal model. For the present case it yields a version of the normal for a population of 2. A systematic way of performing this last step is described in [3].

2.2 Higher Dimensions

The power of this systematic approach becomes apparent as the complexity of the system increases. Imagine, then, that the mass is moving in 3-space, suspended from three orthogonal springs as shown in Figure 7.

The state space is now 6-dimensional, containing three position coordinates (x_1, x_2, x_3) and three velocity coordinates $(\dot{x}_1, \dot{x}_2, \dot{x}_3)$. The total energy is the

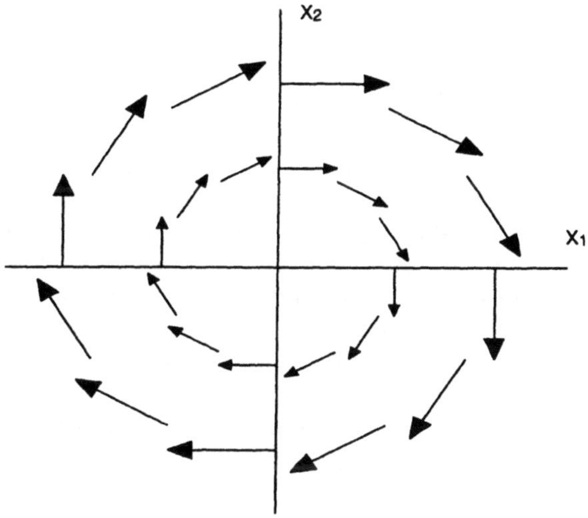

Figure 8 Euclidean 2-space and the vector field generating the rotations (i.e., the Killing field).

sum of the three potential and three kinetic energies:

$$U = \frac{1}{2}k\sum_{i=1}^{3} x_i^2 + \frac{1}{2}m\sum_{i=1}^{3} \dot{x}_i^2.$$

On every $(x_i, \dot{x}_i)$ subspace of the state space we can carry out the same analysis as before. However, we seek a joint distribution on the entire space. The structure of this space is that of a 3-dimensional base space of positions with at every position a 3-dimensional fiber of velocities. Both the position space and the velocity space are Euclidean spaces. That is, it is meaningful to calculate Euclidean norms and angles. This is sometimes stated by saying the Newton's space is isotropic (see Arnold [1]). Another way of saying this is that position space and velocity space are invariant under rotations, as these preserve norms and angles. The rotations can be represented by vector fields that are their infinitesimal generator. This vector field is shown in Figure 8 for a plane.

The isotropicity of space together with the Newtonian structure of the $(x_i, \dot{x}_i)$ gives us vector fields over the entire 6-dimensional state space. The invariant distributions are easily derived from this. One way is to transform coordinates so that the entire space becomes rotational symmetric and then transform back again. As there are now six variables, there are many likelihoods.

Consider first the likelihood for the x_1 alone:

$$p(x_1|U) \propto \frac{1}{\sqrt{U}} \left[1 - \frac{x^2}{2U/k}\right]^{3/2},$$

and compare this to the likelihood in the previous example in Equation (1.2). The only difference is that the value of the exponent is now 3/2 instead of -1/2. This is because the state space is now 6-dimensional. In general for the likelihood of a single variable this exponent is $(N-1)/2-1$ where N is the dimension of the state space. Although this makes no physical sense here, notice that as $N \to \infty$, the likelihood will converge to a normal distribution with mean 0 and variance equal to the average energy.

Consider next the joint likelihood for the first position and velocity :

$$p(x_1, \dot{x}_1|U) \propto \frac{1}{U} \left[1 - \frac{kx_1^2 + m\dot{x}_1^2}{2U}\right]$$

Notice that position and velocity are dependent variables conditional on U. This makes sense in the light of the energy conservation equation in (1.1). The exponent is now 1. In general the exponent is $(N-n)/2-1$, where n is the number of variables before conditioning sign. When we take the imaginary limit with $N \to \infty$, x_1 and $\dot{x}_1$ become independent. The limiting energy consists of an infinite amount of contributions and so the influence of any finite number is negligible.

Other marginal likelihoods can also be determined. For instance, the marginal density for the three positions is:

$$p(x_1, x_2, x_3|U) \propto \left[1 - \frac{k\sum_{i=1}^{3} x_i^2}{2U}\right]^{1/2}.$$

For the three velocities, replace k by m. This distribution is exchangeable. It is also rotationally symmetric. The positions are identically distributed. This is a direct consequence of the isotropy of Euclidean space.

Although they are mathematically closely related, the physical sources of the vector fields in Figure 5 and 8 are entirely distinct. The former is determined by the laws of motion, the latter by the structure of Euclidean space. The former conserves energy, the latter conserves the Euclidean metric. The fact that the former is mathematically closely related to the latter is an peculiarity of Newtonian mechanics, where energy is a quadratic form. The point that may be obscured by this is that the symmetry that applies is completely determined by the physics of the problem, whatever is may be.

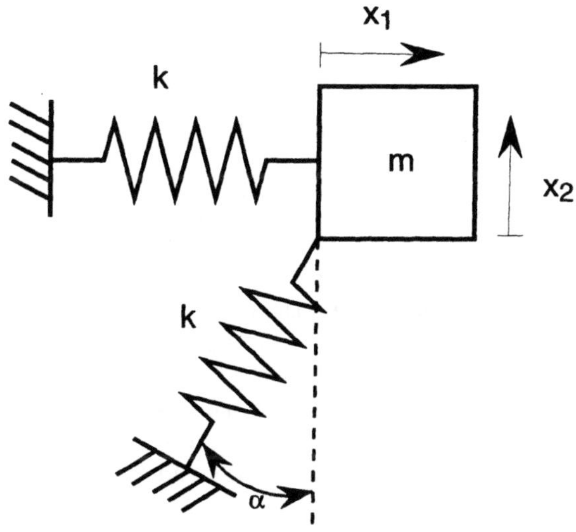

Figure 9 Two dimensional conservative second order system with non-orthogonal springs.

2.3 Dependence

Suppose now that the springs that suspend the mass are not orthogonal as shown in Figure 9. It is interesting to investigate the probabilistic consequences of this change in the physical setup.

Physically what happens is that the coordinates are not independent anymore. A movement along x_1 generates a downward movement along x_2 as the angled spring also extend and generates a downward force on the mass. The potential energy cannot be viewed as the sum of two independent energies but must be calculated as indicated in the following expression for the total energy U:

$$U = \frac{1}{2}k\{x_1 \; x_2\} \begin{bmatrix} 1+\sin^2\alpha & \sin\alpha\cos\alpha \\ \sin\alpha\cos\alpha & \cos^2\alpha \end{bmatrix} \left\{ \begin{array}{c} x_1 \\ x_2 \end{array} \right\} + \frac{1}{2}m\sum_{i=1}^{2}\dot{x}_i^2.$$

Notice that when $\alpha = 0$, we obtain the expression in the previous subsection. The matrix

$$K = k\begin{bmatrix} 1+\sin^2\alpha & \sin\alpha\cos\alpha \\ \sin\alpha\cos\alpha & \cos^2\alpha \end{bmatrix}$$

is known as the stiffness tensor. In the following, let $\vec{x} = \left\{ \begin{array}{c} x_1 \\ x_2 \end{array} \right\}$ for convenience.

Consider now the likelihood for, say, $x_1, x_2, \dot{x}_1$. It is

$$p(x_1, x_2, \dot{x}_1|U) \propto \left[1 - \frac{\vec{x}^T K \vec{x} + m\dot{x}_i^2}{2U}\right]^{-1/2}.$$

Although all the variables are dependent, the dependence among x_1 and x_2 is more fundamental. In fact, recognizing the exponent here to be $(N - n)/2 - 1$ with $N = 4$ and $n = 3$, and letting $N \to \infty$, the velocity becomes independent of the positions in this limit, but the two positions remain dependent. This is the type of dependence that in statistics is usually modelled by multivariate distributions. Here it emerges from the physical dependence of the positions and can clearly be distinguished from the dependence that arises from the finiteness of N.

2.4 Non-linear Systems

The techniques outlined in subsection 2.1 apply equally well to non-linear systems. In fact, for these the state-space need not be diffeomorphic to $\Re^n$, but can be an arbitrary manifold, as will be shown presently with an example. This bears out the advantages of differential-geometric techniques in engineering probability.

Consider the pendulum in Figure 10. Its state space is a cylinder as shown. The position angles ϕ take values on a circle and there is a line of angular velocities $\dot{\phi}$ for each angle. It is also possible to choose the projective coordinates z and $\dot{z}$. The state space remains the same; the projective line on which z takes its values is diffeomorphic to a circle.

Gravity-free: First, assume that the pendulum is in a gravity-free environment, i.e., that $g = 0$ in Figure 10. For instance, it could be spinning horizontally on a table, assuming there is no friction. The total energy is then the kinetic energy of the bob, i.e.,

$$U = \frac{1}{2} m l^2 \dot{\phi}^2,$$

and this is conserved. It follows that the angular velocity is constant,

$$\dot{\phi} = l\sqrt{2U/m},$$

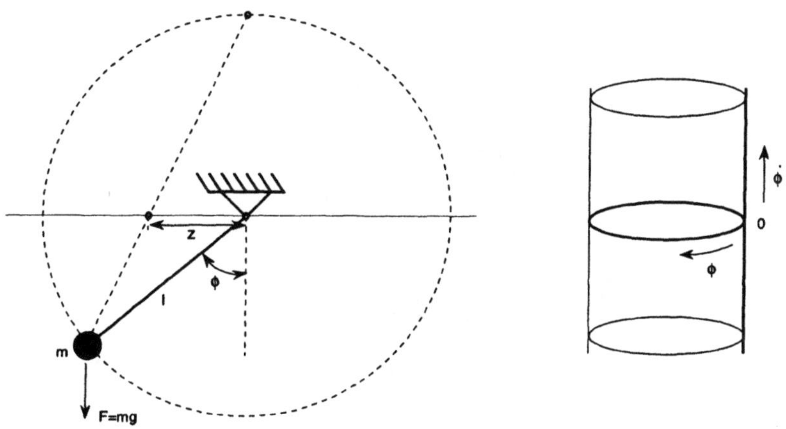

Figure 10 Pendulum (left) and its state space (right) with angular ϕ and angular-velocity $\dot{\phi}$ coordinates; l is the length of the rigid, massless arm and m is the mass of the bob. z is the projective coordinate.

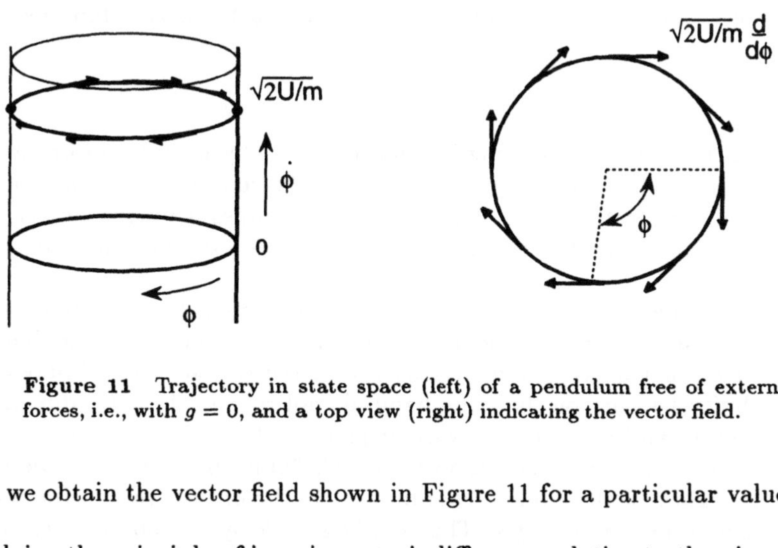

Figure 11 Trajectory in state space (left) of a pendulum free of external forces, i.e., with $g = 0$, and a top view (right) indicating the vector field.

and we obtain the vector field shown in Figure 11 for a particular value of $\dot{\phi}$.

Applying the principle of invariance or indifference relative to the given vector field gives the distribution quantifying a random state of the pendulum in time. In the angle coordinate ϕ this is a uniform density function,

$$p(\phi|U) \propto 1.$$

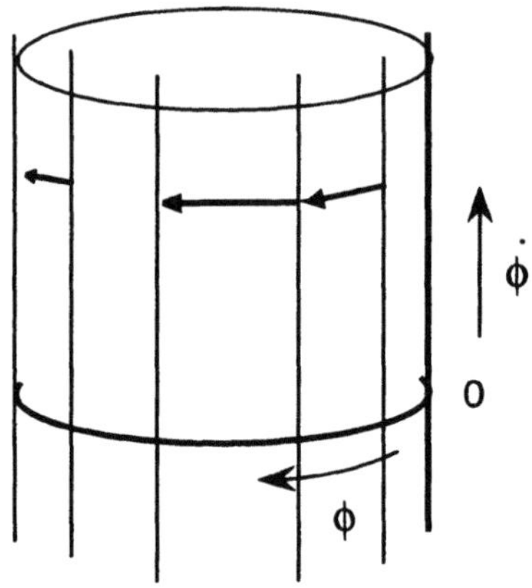

Figure 12 Conditional probability 1-form field for a pendulum swinging in a gravity-free environment. The form lines separate units of 1/10 of probability.

In the projective coordinate z, however, this is a Cauchy density function,

$$p(z|U) \propto \frac{l}{l^2 + z^2}.$$

The point is that there is nothing inherently (i.e., physically) uniform about the distribution. This is only with respect to a choice of coordinates and this choice is arbitrary in this case. The physics determines a unique model, but such a model can be written down in various ways as a density function.

Another problem with the classical density function here is that it is impossible to graph it on the state space. The density axis will be sticking out at a different angle for each point making comparisons among the various values impossible. The probability 1-forms, on the other hand, do not require the extra dimension and give a clean and coordinate invariant picture of the distribution. This is shown in Figure 12. The only reason they look "uniform" is because we have opted in the figure for the ϕ and $\dot{\phi}$ coordinates. In the projective coordinates an entirely different picture emerges. This could be made by appropriate stretching and squeezing the state space manifold so that the projective coordinate line up with the coordinates of the Euclidean space in which the picture lives.

Gravity: Now consider the same pendulum swinging in a gravity field with gravitational constant g. The total energy is now the sum of the potential and the kinetic energy of the bob:

$$U = mgl(1 - \cos\phi) + \frac{1}{2}ml^2\dot{\phi}^2.$$

The possible trajectories pertaining to the various values of U are shown in Figure 13, where the state space cylinder is cut along the $\dot{\phi}$ axis for clarity. For values of $U > 2mgl$ the pendulum can swing around and the trajectory makes a loop around the state space cylinder, attaining all values for ϕ. In the figure these trajectories show up as open curves. For values of $U < 2mgl$ the pendulum swings back and forth, not all values for ϕ are attained, and the trajectories show up as closed curves.

The tangent vector to the trajectory is now:

$$\frac{\sqrt{2U/m - gl(1 - \cos\phi))}}{l}\frac{\partial}{\partial\phi} + mg\sin\phi\frac{\partial}{\partial\dot{\phi}}.$$

Some of these are also shown in Figure 13. Applying the principle of indifference or invariance with respect to this vector field gives the density describing a random state of the system:

$$p(\phi|U) \propto \frac{1}{\sqrt{U}}\left[1 - \frac{1 - \cos\phi}{U/mgl}\right]^{-1/2}.$$

Note that for small values of ϕ, the expression $1 - \cos\phi$ approaches $\phi^2/2$ and this likelihood approaches the likelihood of the position of its linear analog, the spring-mass system, in Expression 1.2. For small deviations, the a pendulum behaves as a linear second-order system. The density function $p(\phi|U)$ given above can be interpreted as a finite-population analog of the normal density on the circle.

The conditional probability 1-forms describing this distribution are shown in Figure 14. For small values of $\dot{\phi}$, the form lines are close to those of the spring-mass system in Figure 6. This is consistent with the result in the previous paragraph. On the other hand, for large values of $\dot{\phi}$ the form lines become like those for the gravity-free pendulum in Figure 12. The reason is as follows: large values of $\dot{\phi}$ imply large values of U. As U increases, the effect of gravity on the motion of the pendulum decreases, and in the limit it spins at a constant rate.

This subsection clearly shows the benefits of the differential-geometric techniques for engineering probability analysis. Most engineering systems are non-linear and their state spaces can be complicated manifolds. Most of the tools

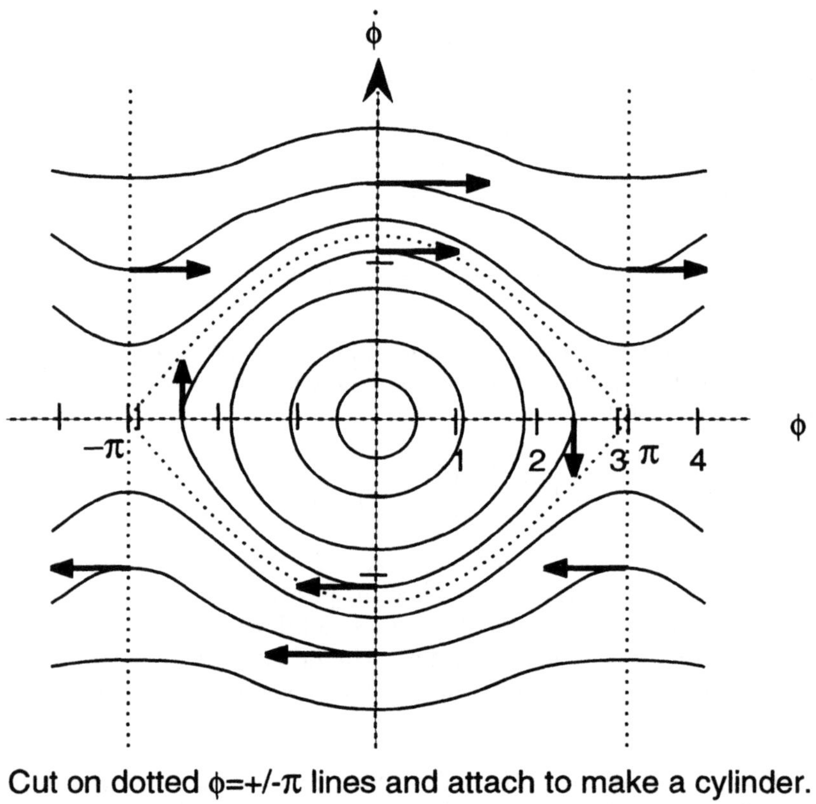

Figure 13 Trajectory in state space of a pendulum in a gravitational field. The lines represent paths of constant energy. Some incidental tangent vector are also shown.

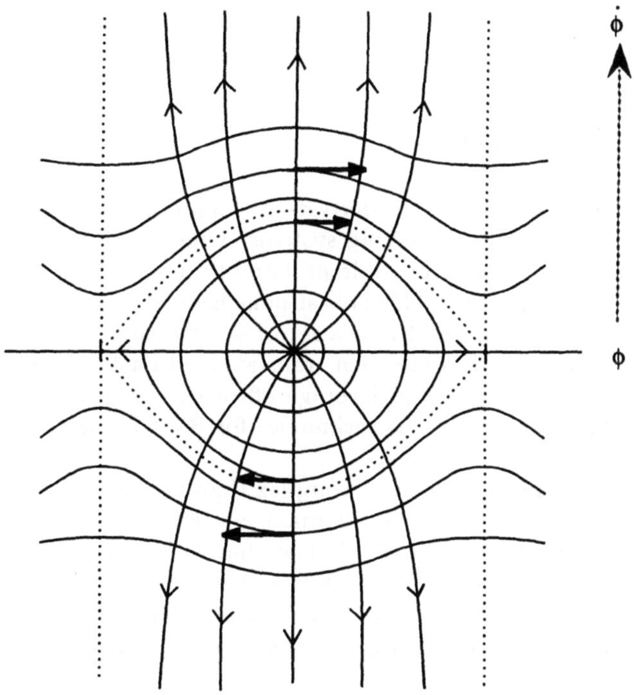

Figure 14 Conditional probability 1-form field for a pendulum swinging in a gravity-free environment. The form lines separate units of 1/5 of probability.

of applied probability, such as density functions or cumulative distribution functions, however, are only useful on $\Re^n$. On the other extreme, the use of measures to represent distributions extends to arbitrary algebras and sigma-algebras, but does not yield a consistent calculus or geometry and there is no guarantee that the solutions are computable. Differential forms only rely on the smoothness of engineering systems, i.e., the fact that most engineering systems can be assumed to be locally linear. As such, they provide a consistent tool of engineering-probability analysis, while guaranteeing computability. In addition, they give very effective graphics.

3 CONCLUSIONS

Engineering probability concerns the derivation of a probability model from the domain expertise regarding engineering systems. One can think of this expertise as constraining the class of possible probability models to a subclass that is consistent with the expertise. This is best done geometrically. The engineering domain expertise is translated into a set of symmetries that constrain the possibly models. These models are then translated into a Bayesian likelihood model. This approach allows us to leverage off of the existing techniques on symmetry and invariance, as well as techniques for Bayesian inference.

Most engineering systems are far more complex than the second order systems considered in this article. Also, the domain expertise often does not have the level of irrefutability that Newton's laws have. Hence, neither will the engineering probability model. A lot of effort will go into identifying reasonable engineering assumptions that then imply a reasonable probability model. This is an important aspect of engineering probability that is not addressed in this article.

There is, however, a place for analyzing simple systems such as the second-order system. Such systems form the basic tools of engineers and it is possible to analyze the dynamic behavior of entire locks or bridges by viewing them as made up of such simple systems. Currently, a common toolbox for engineering probabilistic analysis is lacking and practitioners are forced to start from generalized statistical concepts.

Too often, terms such as "engineering probability" or "engineering statistics" refer to probability or statistics watered-down and made-easy for non-statisticians. This article demonstrates that, on the contrary, engineering probability contains its unique mathematical challenges. The article argued for the usefulness of differential geometry as a common language for engineering and probability: the engineering is modelled by vector fields and the probability by form fields. Differential geometry has the generality to deal with the majority of engineering systems, it provides a calculus and guarantees computability, it gives a coordinate-free representation of the solution which is devoid of any non-physical or ad-hoc assumptions, and it provides for very effective graphics.

REFERENCES

[1] V. I. Arnold. *Mathematical Methods of Classical Mechanics.* Springer-Verlag, New York, 1978.

[2] B. de Finetti. La prevision: ses lois logiques, ses sources subjectives. *Annales de l'Institut Henri Poincare*, 1937. English translation in Kyburg, Jr., H. E. and Smokler, H.E. eds. *Studies in Subjective Probability,* New York: Wiley, 1964.

[3] Max B. Mendel and Peter J. Kempthorne. Operational parameters for the use of bayesian methods in engineering. *REBRAPE, The Brazilian Journal of Probability and Statistics (To appear)*, 1996.

[4] Diaconis P. and Freedman D. A dozen de finetti-style results in search of a theory. *Annales de l'Institut Henri Poincaré*, 23:397–423, 1987.

[5] J. Shortle and M. B. Mendel. A method of drawing probability densities based on the geometry of differential forms. Technical Report ESRC 96-10, University of California at Berkeley, 1996. Engineering Research Systems Center.

PART I

2

OPTIMAL MAINTENANCE DECISIONS FOR THE SEA-BED PROTECTION OF THE EASTERN-SCHELDT BARRIER

Jan van Noortwijk*, Matthijs Kok, and Roger Cooke***

**Department of Mathematics, Delft University of Technology, Delft, The Netherlands*
***Delft Hydraulics, Emmeloord, The Netherlands*

ABSTRACT

To prevent The Netherlands from flooding, a flood defence system has been constructed, which must be inspected and, when needed, repaired. Therefore, one might be interested in obtaining cost-optimal rates of inspection, i.e. rates of inspection for which the expected maintenance costs are minimal and for which the flood defence system is safe.

For optimisation purposes, maintenance models have been developed for two components of the sea-bed protection of the Eastern-Scheldt barrier: (i) the block mats and (ii) the rock dumping. These models enable optimal maintenance decisions to be determined on the basis of (possibly large) uncertainties in the limiting average rates of deterioration. The modelling assumption that the stochastic processes of scour erosion and rock displacement depend just on limiting averages leads us to regard them as generalised gamma processes.

Keywords. maintenance, gamma processes, renewal theory, decision theory, Eastern-Scheldt barrier, rock displacement, scour erosion.

R. Cooke et al. (eds.), Engineering Probabilistic Design and Maintenance for Flood Protection, 25–56.

1 INTRODUCTION

In this chapter, we consider the problem of inspecting the sea-bed protection of the Eastern-Scheldt storm-surge barrier. Since the barrier is planned to function for a period of 200 years, it is inspected to reveal possible deterioration that might endanger the stability of the barrier. Therefore, one might be interested in obtaining cost-optimal rates of inspection, i.e. rates of inspection for which the expected maintenance costs are minimal and for which the barrier is safe.

A large number of papers have been published on the subject of optimising maintenance through mathematical models. For an, inherently incomplete, overview see Barlow & Proschan [2], McCall [19], Pierskalla & Volker [21], Sherif & Smith [24], Sherif [25], Valdez-Flores & Feldman [27], and Cho & Parlar [4]. Most maintenance optimisation models are based on lifetime distributions or Markovian deterioration models. Unfortunately, only a few of them have been applied (see Dekker [8]). According to De Jonge, Kok & Van Noortwijk [6], there are two possible reasons for this poor applicability. First, from the theoretical point of view, there is often no interest in "details" that are of practical importance: a problem description is often lacking or even purely hypothetical. Second, from the practical point of view, there is little experience in using maintenance optimisation models and it is often hard to gather data for estimating either the parameters of a lifetime distribution or the transition probabilities of a Markov chain. Moreover, in case of well-planned preventive maintenance, complete lifetimes will be observed rarely. The authors hope to develop a methodology that might bridge the gap between theory and practice by modelling maintenance on the basis of the main uncertainties involved: the values of the limiting average rates of deterioration. To achieve this, deterioration processes can best be regarded as generalised gamma processes. In The Netherlands, generalised gamma processes have also been used to model decision problems for optimising maintenance of beaches, berm breakwaters, and dykes (see Van Noortwijk et al. [29, 30] and Speijker et al. [26]).

The chapter is organised as follows. A brief description on the Eastern-Scheldt barrier is given in Sec. 2. In Secs. 3 and 4, we present maintenance models for two components of the barrier: the block mats and the rock dumping, respectively. Some necessary definitions and theorems are presented in two appendices.

2 THE EASTERN-SCHELDT BARRIER

With storm-induced tides of some 4 metres above average sea level, the flood of February 1, 1953, caused a severe catastrophe in Zeeland, The Netherlands. Almost 200,000 hectares of polderland flooded, resulting in huge losses of life and property. In the south-west of The Netherlands, 1,835 people and tens of thousands of animals were drowned. To avoid future losses due to floods like the one in 1953, the Dutch parliament adopted the so-called Delta Plan. The greater part of this plan called for raising the dykes and for closing the main tidal estuaries and inlets by a network of dams and barriers. Since the Delta Plan will soon be completed, the attention is shifting from building structures to maintaining structures. Hence, the use of maintenance optimisation models is of considerable interest.

This chapter is devoted to modelling preventive maintenance of the most expensive and the most complicated structure of the Delta Works: the Eastern-Scheldt storm-surge barrier. The design of this lift-gate barrier is complex for it has to satisfy requirements in the following areas: (i) safety (flood protection during severe storm-surges when the gates are closed), (ii) environment (preservation of the natural salt-water environment during normal weather and hydraulic conditions when the gates are open), and (iii) transport (shipping access to the North-Sea and a road-connection).

The Eastern-Scheldt barrier has been built in three closure gaps separated by two artificial islands (see Fig. 1). It has 62 pier-supported steel gates each with a span of nearly 42 metres and a height varying from 6 to 12 metres. To provide for the long-term stability of the barrier, the supporting concrete piers are embedded with several layers of rock and an adjoining sea-bed protection has been constructed with a width of about 500 metres on either side of the center line of the barrier. This sea-bed protection consists of asphalt mastic and block mats in the outer periphery, and graded-filter mattresses under the piers (see Fig. 2). Since the protection can be damaged, it is monitored for the appearence of scour holes. In this situation, the rates of inspection and the costs of maintenance have to be optimised. For brief summaries on the technical aspects and the maintenance aspect of the Eastern-Scheldt barrier, see Rijkswaterstaat [22] and Watson & Finkl [31], and De Jonge, Kok & Van Noortwijk [6], respectively.

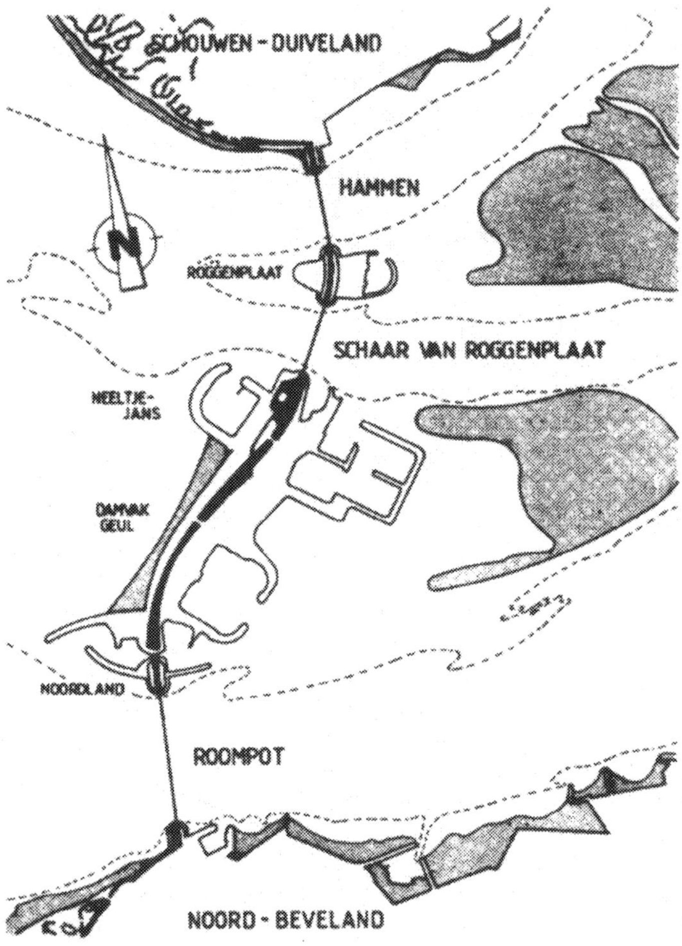

Figure 1 The map of the Eastern-Scheldt estuary showing the two artificial dredge-improved islands ('Roggenplaat' and 'Neeltje Jans') and the three tidal channels controlled by lift-gate barriers ('Hammen', 'Schaar', and 'Roompot').

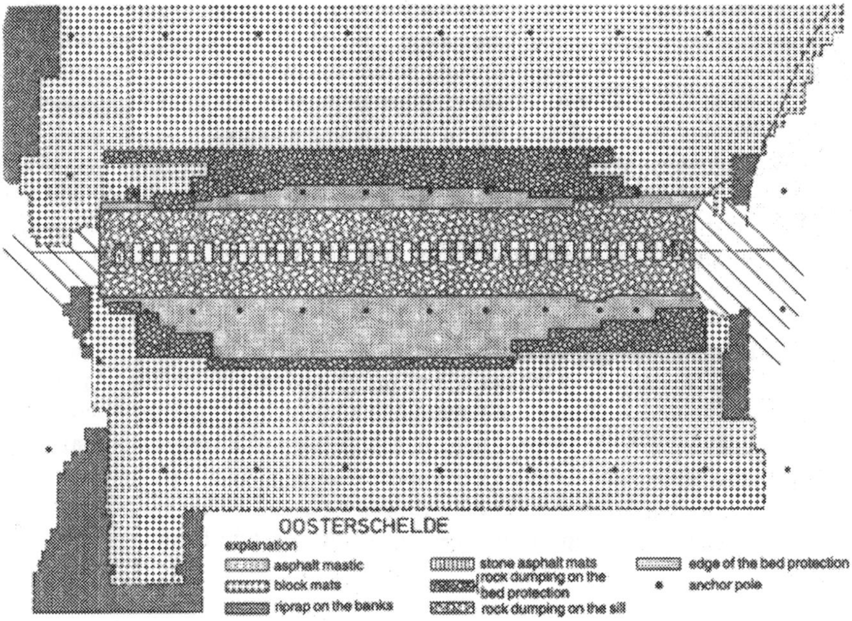

Figure 2 View from above of the sea-bed protection of the Eastern-Scheldt barrier at the tidal channel 'Roompot' (from Rijkswaterstaat (1994)).

3 MAINTENANCE OF THE BLOCK MATS

The block mats consist of synthetic material to which small concrete-blocks (with a height of 17 cm) are attached in a regular pattern. The purpose of this section is to obtain safe and cost-optimal rates of inspection for these mats.

3.1 Inspection and repair of scour holes

Due to extreme tidal currents or ship anchorings, the block mats may be damaged in such a way that sands wash away and scour holes appear. To detect possible scour, the block mats are inspected by means of acoustic measurements. If acoustic inspection reveals a scour hole, then a visual dive inspection will be carried out, followed by a repair. Scour holes can only be detected when they are deeper than about 2 metres. By approximation, we therefore assume the probability of detection to be equal to one when the scour hole is deeper than $z = 2$ and zero otherwise, where z is the detectability level. To confirm the statement that there is often a lack of data in practice: up to now, no scour holes have been detected!

Since preventive maintenance is based on the condition of the block mats, we are dealing with so-called condition-based preventive maintenance. Apart from condition-based repairs, it might be economic to perform (aperiodic) condition-based inspections as well. In practice, however, periodic inspections are often to be preferred since the necessary manpower and budget can be anticipated and scheduled well beforehand. Furthermore, we assume that inspection of the whole block mats takes negligible time, does not degrade the block mats, and entails a cost c_I.

A repair is defined as placing graded rip-rap on a scour-hole surface approximately being a hemisphere of radius h, where h is the scour hole depth (in metres). The costs of repairing one scour hole can be subdivided into the fixed cost c_f (cost of mobilisation like shipping transport) and the variable cost c_v (cost per square metre rip-rap). Hence, the costs of repairing one scour hole, which is h metres deep, are

$$c(h) = c_f + 2\pi h^2 c_v. \tag{2.1}$$

Although we assume possible repairs to be carried out during inspections, the decision model may be extended with delay-times between detecting holes and repairing holes.

In summary, the block mats must be inspected to avoid instability of the barrier due to the following uncertain deterioration characteristics: (i) the average rate of occurrence of scour holes at the whole block mats and (ii) the average rate of current-induced scour erosion given there is a scour hole. Because there is no deterioration data available, we have to rely on prior expert judgment. Moreover, since the decision problem of obtaining safe and cost-optimal inspection rates is characterised by the above two uncertainties, the decision model should be based on them as well. These uncertainties can best be represented by probability distributions, where Bayes' theorem can be used to update subjective prior opinion with actual observations.

3.2 The rate of occurrence of scour holes

The lack of observations has prompted us to assume that the rate of occurrence of scour holes does not depend on the location and the time. Although this assumption may be criticised for being unrealistic, the probabilistic models representing location- and time-dependent inter-occurrence times are inherently more complex and, not the least important, require many observations to estimate their parameters. Recall that no scour holes have been observed yet and that, as a consequence, even subjective opinion on *average* rates is hard to obtain.

Given the above problem description, we shall derive the probability model of the scour-hole inter-occurrence times by making two reasonable judgments: the inter-occurrence times (i) are exchangeable and (ii) exhibit the "lack of memory" property. These are explained below. Let us denote the successive times between occurrences of scour holes by the infinite sequence of non-negative real-valued random quantities $T_1, T_2, \ldots$

First, the inter-occurrence times are assumed to be *exchangeable*: i.e. the order in which the scour holes occur is judged to be irrelevant. In mathematical terms, this can be interpreted as that the probability density function of the random vector $\mathbf{T}_n = (T_1, \ldots, T_n)$ is invariant under all $n!$ permutations of the coordinates, i.e.

$$p_{T_1,\ldots,T_n}(t_1, \ldots, t_n) = p_{T_1,\ldots,T_n}\left(t_{\pi(1)}, \ldots, t_{\pi(n)}\right), \tag{2.2}$$

where π is any permutation of $1, \ldots, n$. The infinite sequence of random quantities $\{T_i : i \in \mathbb{N}\}$ is said to be exchangeable if $\mathbf{T}_n$ is exchangeable for each $n \in \mathbb{N}$. The assumption of exchangeability is weaker than the assumption of independence.

Second, the inter-occurrence times are assumed to exhibit the *"lack of memory" property*: i.e. the probability distribution of the remaining time until the occurrence of the first scour hole does not depend on the fact that no scour hole has appeared yet since the completion of the barrier in 1986 (for a formal definition, see Theorem 2 of the appendix). Another explanation of the "lack of memory" property is the following. Suppose the second scour hole has occurred at time t, i.e. $t_1 + t_2 = t$. If the occurrence time of the first scour hole, t_1, could with equal probability be any time in the interval $[0, t]$, then the "lack of memory" property holds. Thus, $p_{T_1|T_1+T_2}(t_1|\, t) = t^{-1} I_{[0,t]}(t_1)$, being the uniform distribution on $[0, t]$.

Under the assumptions that the infinite sequence $T_1, T_2, \ldots$ is exchangeable and satisfies the "lack of memory" property for all $n \in \mathbb{N}$, we can write the joint probability density function of $T_1, \ldots, T_n$ as a mixture of conditionally independent exponentials (using Theorem 2 or 3 from the appendix):

$$p_{T_1,\ldots,T_n}(t_1, \ldots, t_n) = \int_0^\infty \prod_{i=1}^n \frac{1}{\lambda} \exp\left\{-\frac{t_i}{\lambda}\right\} dQ_\Lambda(\lambda) = f_n(\textstyle\sum_{i=1}^n t_i) \qquad (2.3)$$

for $(t_1, \ldots, t_n) \in \mathbb{R}_+^n$ and zero otherwise, where $\mathbb{R}_+ = [0, \infty)$. The infinite sequence of random quantities $\{T_i : i \in \mathbb{N}\}$ is said to be l_1-*isotropic* (or l_1-norm symmetric), since its distribution can be written as a function of the l_1-norm. In general, an infinite sequence of random quantities is said to be l_p-*isotropic*, if its distribution can be written as a function of the l_p-norm (see Misiewicz & Cooke [20]). The random quantity Λ, with probability distribution Q_Λ, describes the uncertainty about the limiting average inter-occurrence time of scour holes: $\lim_{n\to\infty}[(\sum_{i=1}^n T_i)/n]$ (see e.g. Barlow & Mendel [1] and Van Noortwijk, Cooke & Kok [28]). Note that the information about the unknown parameter λ, contained in $T_1, \ldots, T_n$, is summarised by the statistic $[n, \sum_{i=1}^n T_i]$ which is sufficient for λ. The characterisation of Eq. (2.3) in terms of the "lack of memory" property is due to Diaconis & Freedman [10] (see Theorem 2 in the appendix) and the characterisation of Eq. (2.3) in terms of conditioning on the sufficient statistic $[n, \sum_{i=1}^n T_i]$ is due to Diaconis & Freedman [9] (see Theorem 3 in the appendix). For an overview on statistical modelling using exchangeability and sufficiency, see Bernardo & Smith [3, Ch. 4].

For modelling the occurrences of scour holes, only the probability distribution of the average inter-occurrence time remains to be determined. To keep the mathematics of the decision model tractable, we impose the property of *posterior linearity* introduced by Diaconis & Ylvisaker [11, 12], i.e. $E(T_2|\, T_1 = t_1) = c_1 t_1 + c_2$ for some constants $c_1, c_2 > 0$. Remark that, due to exchangeability, before observing T_1, $E(T_2) = E(T_1)$. If posterior linearity holds, then the mixing distribution Q_Λ is the inverted gamma distribution $\text{Ig}(\lambda|\alpha,\beta)$ (see also Theorem 2 from the appendix). The mathematical tractability is especially useful if one wants to update the prior distribution $\text{Ig}(\lambda|\alpha,\beta)$ with actual observations $t_1,\ldots,t_n$. In fact, using Bayes' theorem, the posterior distribution is $\text{Ig}(\lambda|\alpha+n,\beta+\sum_{i=1}^{n} t_i)$. Owing to the fact that the posterior mean can be written as a linear combination of the prior mean and the sample mean, the property of posterior linearity has been satisfied.

3.3 The rate of current-induced scour erosion

Beside the uncertainty in the average rate of occurrence of scour holes, we have to take into account the uncertainty in the average rate of current-induced scour erosion. This average is taken over all possible scour holes deeper than z metres and over all possible locations at the block mats. For each scour hole, erosion is measured in terms of the scour-hole depth h, where $h > z$ and $z = 2$.

The purpose of this subsection is to characterise the stochastic process of scour erosion in terms of the only (subjective) information that is available: the probability distribution of the average rate of scour erosion. In doing so, we shall adopt the following two assumptions: with respect to any uniform time-partition and any location at the block mats, the increments of erosion (i) are non-negative and exchangeable; and (ii) have a joint conditional probability distribution, given their sum, which can be represented by a multi-dimensional beta distribution (Dirichlet distribution). Let us denote the erosion process by $\{X(t) : t \geq 0\}$: a non-decreasing continuous-time stochastic process with $\Pr\{X(0) = 0\} = 1$ and $X(t)$ representing the cumulative erosion, per scour hole, at time t.

The first assumption means: for every uniform time-partition in time-intervals of length $\tau > 0$, the infinite sequence of random increments of erosion, $D_i(\tau) = X(i\tau) - X([i-1]\tau)$, $i \in \mathbb{N}$, is assumed to be exchangeable, where $D_i(\tau) \geq 0$ for all i.

The second assumption means: for all $\tau > 0$, the conditional probability density function of the first increment of erosion, when the sum of the first and the second increment is given, can be expressed as a transformed beta distribution with both parameters equal to $a\tau$, i.e.

$$p_{D_1(\tau)|X(2\tau)}(\delta_1|x) =$$

$$= \frac{\Gamma(2a\tau)}{[\Gamma(a\tau)]^2} \frac{\delta_1^{a\tau-1}[x-\delta_1]^{a\tau-1}}{x^{2a\tau-1}} I_{[0,x]}(\delta_1) = \text{Be}\left(\frac{\delta_1}{x}\middle| a\tau, a\tau\right)\frac{1}{x} \qquad (2.4)$$

for some constant $a > 0$ with

$$\begin{aligned} E(D_1(\tau)|X(2\tau)=x) &= x/2, \\ \text{Var}(D_1(\tau)|X(2\tau)=x) &= [x/2]^2/(2a\tau+1). \end{aligned}$$

We now indicate how Eq. (2.4) is derived. To begin with, if D_1 were not symmetrically distributed about its mean, $x/2$, then the random quantities D_1 and D_2 would not be exchangeable. Hence, the parameters of the beta distribution should be equal. Moreover, we can actually derive Eq. (2.4) by conditioning on sums of increments. This characterisation extends the results of Diaconis & Freedman [9] by achieving consistency in the sense that probability distributions of increments and sums of increments belong to the same family of distributions and by assuming the probability model to be independent of the scale of measurement (i.e. to be a scale mixture). For the mathematics of the characterisation in terms of conditioning on sums of increments, we refer to Theorem 4 from the appendix.

By characterising exchangeable erosion processes in terms of conditional distributions given sums of increments, i.e. in terms of Eq. (2.4), for all $\tau > 0$, we so characterise the *generalised gamma process*. Indeed, it follows from Theorem 3 (see the appendix) with $h(y) = y^{a\tau-1}/\Gamma(a\tau)$ that, for all $\tau > 0$, the joint probability density function of the increments $D_1(\tau), \ldots, D_n(\tau)$ can be written as a mixture of conditionally independent gamma densities:

$$\begin{aligned} & p_{D_1(\tau),\ldots,D_n(\tau)}(\delta_1,\ldots,\delta_n) = \\ & = \int_0^\infty \prod_{i=1}^n \frac{\delta_i^{a\tau-1}}{\Gamma(a\tau)} \left[\frac{a\tau}{\theta}\right]^{a\tau} \exp\left\{-\frac{a\tau\delta_i}{\theta}\right\} dP_{\Theta(\tau)}(\theta) \qquad (2.5) \end{aligned}$$

for some constant $a > 0$ with

$$\begin{aligned} E(X(n\tau)) &= E(n\Theta(\tau)), \\ \text{Var}(X(n\tau)) &= \left[1 + \frac{1}{na\tau}\right] E([n\Theta(\tau)]^2) - [E(n\Theta(\tau))]^2 \end{aligned}$$

for all $\tau > 0$, provided the first and the second moment of the probability distribution of $\Theta(\tau)$ exist.

The generalised gamma process has three useful properties.

First, the probability distribution $P_{\Theta(\tau)}$ on the random quantity $\Theta(\tau)$, with possible values $\theta \in (0, \infty)$, represents the uncertainty in the unknown limiting average amount of erosion per time-interval of length τ: $\lim_{n\to\infty}[(\sum_{i=1}^{n} D_i(\tau))/n]$. By the strong law of large numbers for exchangeable random quantities, the average converges with probability one if $E(D_1(\tau)) < \infty$ (see Chow & Teicher [5, p. 227]).

Second, the summarisation of the n random quantities $D_1(\tau), \ldots, D_n(\tau)$ in terms of the statistic $[n, \sum_{i=1}^{n} D_i(\tau)]$ is sufficient for the unknown limiting average erosion $\Theta(\tau)$. In fact, the characterisation in terms of conditioning on sums of random quantities is motivated by sufficiency ideas, since, by sufficiency, the conditional probability density function $p_{D_1(\tau)|X(2\tau),\Theta(\tau)}(\delta_1 | x, \theta)$ does not depend on θ.

Third, the mixture of gamma's in Eq. (2.5) transforms into a mixture of exponentials if $\tau = a^{-1}$. As we shall see in Secs. 3 and 4, for this unit-time length, denoted by $\Delta = a^{-1}$, many probabilistic properties of the stochastic process, like the probability of exceedence of a failure level, can be expressed in explicit form conditional on the limiting average. The unit time for which the increments of erosion are distributed according to a mixture of exponentials, i.e. are l_1-isotropic, follows directly from the characterisation in terms of conditioning on sums of increments in Eq. (2.4). From this equation (for fixed $\tau > 0$), it can be seen that the smaller the unit-time length for which the increments are l_1-isotropic, i.e. the smaller $\Delta = a^{-1}$, the more deterministic the erosion process.

Note that specifying the l_1-isotropic grid of the gamma process is similar to specifying the variance of the Brownian motion with drift (see e.g. Karlin & Taylor [15, Ch. 7]). This stochastic process is often used to model stochastic deterioration. Unfortunately, the Brownian motion allows for a probability of "negative deterioration", especially when large uncertainties are involved.

In conclusion, we advocate regarding the stochastic erosion process as a generalised gamma process with probability distribution on the limiting average rate of erosion. This process does not entail extra difficulties, but has the advantage that it models realistic non-negative deterioration rather than unrealistic real-valued deterioration. As for the limiting average inter-occurrence time, we impose the property of posterior linearity and assume the probability distribution of the limiting average erosion, per unit time of length Δ, to be the inverted gamma distribution $\mathrm{Ig}(\nu, \mu)$.

3.4 The maintenance decision model

Based on the two stochastic processes of the occurrences and the development of scour holes, which are judged to be independent, we can formulate the maintenance optimisation model. From now on, we choose our units of time so that the increments of scour erosion per unit time are l_1-isotropic (with respect to $\{([n-1]\Delta, n\Delta] : n \in \mathbb{N}\}$). For notational convenience, let $D_n = D_n(\Delta)$, $X_n = \sum_{h=1}^{n} D_h$ for all $n \in \mathbb{N}$, and let Θ represent the uncertainty in the limiting average rate of scour erosion $\lim_{n\to\infty}[(\sum_{i=1}^{n} D_i)/n]$. Subsequently, we determine the expected number of scour holes that occur per unit time and the expected cumulative amount of detectable scour erosion.

When the limiting average inter-occurrence time has the value λ, then the scour holes arrive according to a Poisson process with arrival rate λ^{-1}. It is well-known (see e.g. Karlin & Taylor [15, pp. 173-175]) that the number of scour holes occurring in unit time i follows a Poisson distribution with parameter Δ/λ:

$$\Pr\{\text{number of holes in } i\text{th unit time} = j \mid \lambda\} = \frac{(\Delta/\lambda)^j e^{-\Delta/\lambda}}{j!} \tag{2.6}$$

for $j = 0, 1, 2, \ldots$. The expected number of scour holes that occur in unit time i can simply be written as $s_i(\lambda) = \Delta/\lambda$ and does not depend on i (due to the "lack of memory" property).

When the limiting average rate of scour erosion per unit time has the value θ, then a scour hole that first could have been detected in unit time i, but is inspected in unit time k, entails the following expected costs of repair (using Eq. (2.1)):

$$u_i(\theta, k) = c_f + 2\pi c_v E\left([z + X_{k-i+1}]^2 \mid \theta\right) \tag{2.7}$$

where $1 \le i \le k$ and X_n having a gamma distribution $\mathrm{Ga}(n, 1/\theta)$ for all $n \in \mathbb{N}$ (for the definition of the gamma distribution, see the appendix).

Our main interest is to determine an inspection interval of length $k\Delta$, $k \in \mathbb{N}$, for which the expected maintenance costs are minimal and the barrier is safe, where inspections are carried out at times $\{jk\Delta : j \in \mathbb{N}\}$. Let $L(\lambda, \theta, k)$ be the monetary *loss* when the decision-maker chooses inspection interval k, $k \in \mathbb{N}$, and when the limiting averages λ and θ are given. Under the requirement that the barrier is safe, the decision-maker can best choose the inspection interval k^* whose expected loss, $E(L(\Lambda, \Theta, k^*))$, is minimal. The decision k^* is called an *optimal decision* (see e.g. DeGroot [7, Ch. 8]).

The best choice for the loss is the expected *average costs per year* (see Sec. 4), which can be determined by averaging the maintenance costs over an unbounded time-horizon. Since we have renewal cycles of length $k\Delta$, the average costs per year becomes:

$$L(\lambda, \theta, k) = \frac{c_I + \sum_{i=1}^{k} u_i(\theta, k) s_i(\lambda)}{k\Delta}, \tag{2.8}$$

where the numerator consists of the costs of inspection c_I plus the costs of repairing k possible scour holes, summed over all units of time in which they may occur, times the expected number of occurrences. Note that since the expected discounted costs over an unbounded horizon (see Sec. 4) approach zero, from above, for unbounded inspection-intervals, the criterion of discounted costs is not useful for solving this decision problem.

The evaluation of the average costs in Eq. (2.8) is straightforward: by induction and using the gamma integral, we get

$$\sum_{i=1}^{k} E\left(X_{k-i+1}^{m} \,|\, \theta\right) = m!\,\theta^{m} \sum_{i=1}^{k} \binom{m+k-i}{k-i} = m!\,\theta^{m} \binom{m+k}{m+1} \tag{2.9}$$

for $m = 0, 1, 2, \ldots$ and $k = 1, 2, \ldots$. Substitution of Eqs. (2.7) and (2.9) into Eq. (2.8) yields

$$L(\lambda, \theta, k) = \tag{2.10}$$
$$= \frac{c_I}{k\Delta} + \left\{c_f + 2\pi c_v \left[z^2 + z\theta(k+1) + \theta^2 (k+1)(k+2)/3\right]\right\} \frac{1}{\lambda}.$$

Up to now, we did not incorparate the possibility of a severe failure of the block mats such that the barrier is unsafe. Strictly speaking, costs of failure due to potential unsafe situations should be incorporated as well. However, what is "unsafe" and what are the costs of failure involved? The costs of failure not only consist of costs due to damaged block mats, but also of possible costs due to instability of the barrier, and, when there is a severe storm-surge, of possible costs due to flooding. Unfortunately, these costs are very hard to determine or to assess. For this reason, we have chosen to leave out the failure costs, but to introduce an upper bound for the inspection interval with the following property: when this upperbound is crossed, the block mats are said to be unsafe in the sense that there is at least one scour hole deeper than a certain failure level, say y. The probability of this event should be smaller than a predefined norm probability which itself is a function of the inspection-interval length $k\Delta$. For example: $1 - (1 - p_{norm})^{k\Delta}$, where p_{norm} is the annual norm probability.

In mathematical terms, the probability of failure of the block mats can be expressed as follows. By assuming the scour holes being independent given θ and by rewriting the probability of the event "in $(0, k\Delta]$ at least one scour hole occurs that is deeper than y metres" as one minus the probability of the event "in $(0, k\Delta]$ no scour holes occur that are deeper than y metres", we get

$$\begin{aligned} v_k(\lambda, \theta) &= \Pr\{\text{in } (0, k\Delta] \text{ at least one hole is deeper than } y \mid \lambda, \theta\} \\ &= 1 - \prod_{i=1}^{k} \sum_{j=0}^{\infty} \frac{(\Delta/\lambda)^j e^{-\Delta/\lambda}}{j!} \left[\Pr\{z + X_{k-i+1} \leq y \mid \theta\}\right]^j \\ &= 1 - \exp\left\{-\frac{\Delta}{\lambda} \sum_{i=1}^{k} \sum_{h=1}^{k-i+1} \Pr\{X_{h-1} \leq y - z, X_h > y - z \mid \theta\}\right\}, \end{aligned}$$

where the probability of failure of one scour hole follows the Poisson distribution:

$$\Pr\{X_{h-1} \leq x, X_h > x \mid \theta\} = \frac{1}{(h-1)!} \left[\frac{x}{\theta}\right]^{h-1} \exp\left\{-\frac{x}{\theta}\right\} = q_h(\theta, x) \qquad (2.11)$$

for $h = 1, 2, \ldots$, $\theta > 0$, and $x = y - z$ (see Van Noortwijk, Cooke & Kok [28]).

In summary, the decision-maker can best choose the inspection-interval k^* whose expected long-term average costs of maintenance are minimal and whose expected probability of failure is safe:

$$\begin{aligned} E(L(\Lambda, \Theta, k^*)) &= \min_{k \in \mathcal{D}} E(L(\Lambda, \Theta, k)), \text{ where} \qquad (2.12) \\ \mathcal{D} &= \left\{k : k \in \mathbb{N}; E(v_k(\Lambda, \Theta)) < 1 - (1 - p_{norm})^{k\Delta}\right\}. \end{aligned}$$

Although explicit computation of $E(v_k(\Lambda, \Theta))$ is not possible, reasonably sharp lower and upper bounds have been found (see Theorem 1 of the appendix).

For obtaining optimal inspection and repair decisions for the block mats, we use the parameters in Table 1. The unit time for which the increments of scour erosion are distributed as mixtures of exponentials has been determined by specifying the conditional probability density function of the amount of scour erosion in a period of six months when the amount of erosion in a period of one year is given to be 10 metres (using Eq. (2.4) shown in Fig. 3). The optimal decision k^*, satisfying Eq. (2.12), is an inspection interval of one year whose expected average costs per year are 2×10^5 Dutch guilders. Although the maintenance costs are minimal for an inspection interval of 1.5 years (see Fig. 4), the barrier is unsafe for inspection intervals larger than 1 year (see Fig. 5). For practical purposes, no less important than obtaining a unique optimal decision, however, is obtaining a range of nearly cost-optimal and safe decisions. The decision-maker can find an optimum balance between cost and safety using the curves in Figs. 4 and 5.

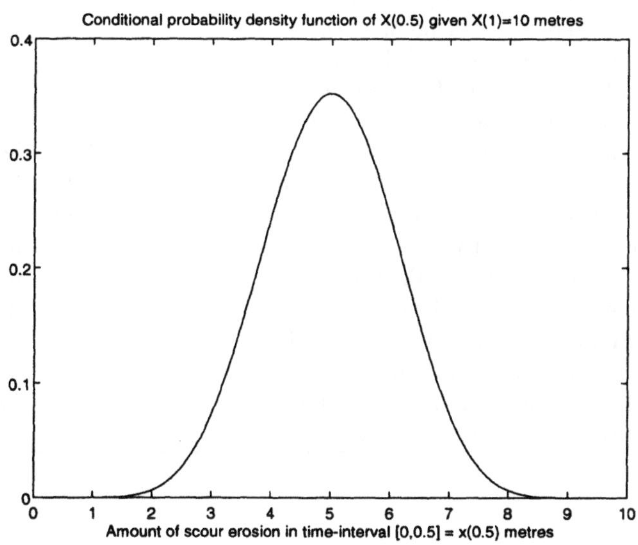

Figure 3 The conditional probability density function of the amount of scour erosion in a period of six months, $X(0.5)$, when the amount of erosion in a period of one year is $X(1) = 10$ metres.

Table 1 The parameters of the maintenance model for the block mats.

parameter	description	value	dimension
Δ	unit time	0.05	year
Θ	average rate of scour erosion	$(0,\infty)$	m/unit time
$\theta_{0.05}/\Delta$	[5%-percentile of $\text{Ig}(\theta\|\nu,\mu)$]/$\Delta$	7	m/year
$\theta_{0.95}/\Delta$	[95%-percentile of $\text{Ig}(\theta\|\nu,\mu)$]/$\Delta$	13	m/year
ν	shape parameter of $\text{Ig}(\theta\|\nu,\mu)$	28.7	
μ	scale parameter of $\text{Ig}(\theta\|\nu,\mu)$	13.3	
$E(\Theta)$	expected average rate of scour erosion	0.5	m/unit time
$E(\Theta/\Delta)$	expected average rate of scour erosion	10	m/year
Λ	average scour-hole inter-occurrence time	$(0,\infty)$	year
$\lambda_{0.05}$	5%-percentile of $\text{Ig}(\lambda\|\alpha,\beta)$	1	year
$\lambda_{0.95}$	95%-percentile of $\text{Ig}(\lambda\|\alpha,\beta)$	10	year
α	shape parameter of $\text{Ig}(\lambda\|\alpha,\beta)$	2.46	
β	scale parameter of $\text{Ig}(\lambda\|\alpha,\beta)$	5.46	
$E(\Lambda)$	expected average inter-occurrence time	4	year
c_I	costs of inspection	$125,000$	Dfl
c_f	fixed costs of repairing one scour hole	$100,000$	Dfl
c_v	variable cost of rip-rap	159	Dfl/m^2
h	scour hole depth	$(0,\infty)$	m
z	scour-hole detectability level	2	m
y	scour-hole failure level	15	m
p_{norm}	annual norm probability of failure	0.01	
k	inspection-interval length	$\mathbb{N}$	unit time
k^*	optimal inspection-interval length	20	unit time
$k^*\Delta$	optimal inspection-interval length	1	year

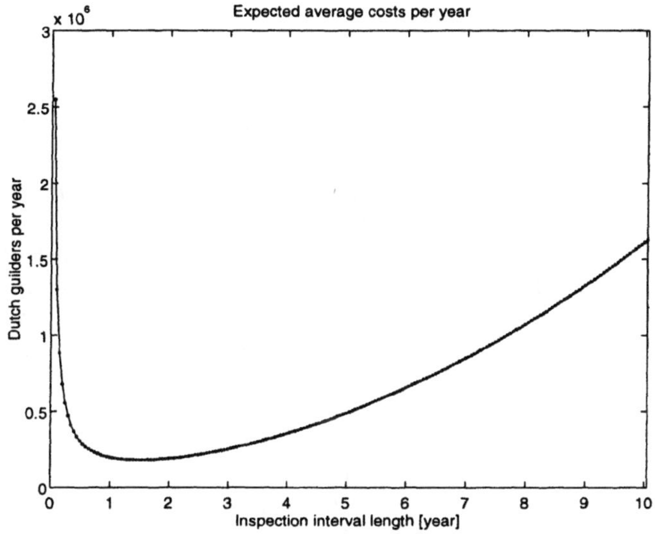

Figure 4 The expected average costs per year for the block mats.

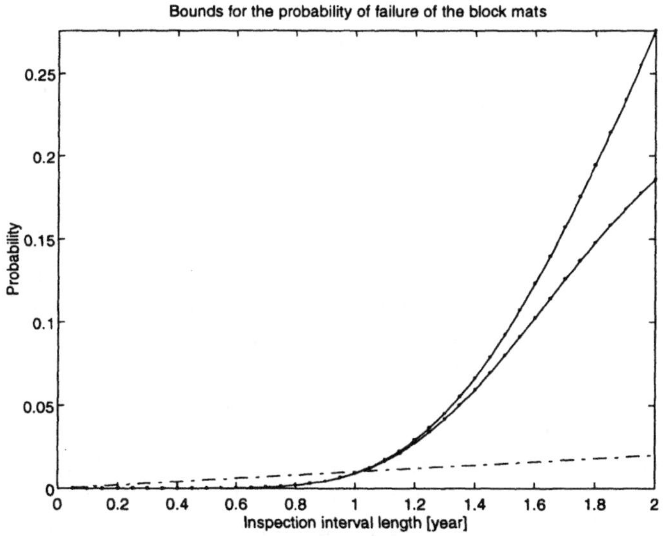

Figure 5 The lower and upper bounds for the expected probability of failure of the block mats crossed by the norm probability of failure.

4 MAINTENANCE OF THE ROCK DUMPING

Using a similar method to that used for the block mats, we can obtain cost-optimal rates of inspection for the rock dumping of the barrier. Millions of tons of rock rubble were placed at the sea-bed protection near the center line of the barrier (see Fig. 2). This protection is subject to current-induced rock displacement, which has to be monitored by means of acoustic measurements and, if necessary, has to be repaired. The inspection problem is due to Kok [16, 17]; it also has been studied by Van Noortwijk, Cooke & Kok [28]. Because it was not their purpose to determine optimal maintenance decisions, we revisit their inspection problem and use their failure model to determine optimal rates of inspection.

By the same reasoning as for the process of scour erosion, we can best regard the stochastic process of rock displacement as a generalised gamma process with an inverted gamma distribution as mixing measure (see Sec. 3.3). The infinite sequence of l_1-isotropic increments of rock displacement is denoted by $\{D_i : i \in \mathbb{N}\}$. The resistance of the upper rock layer of the rock dumping, R, is defined as the number of stones removed (at time zero: $r_0 = 0$). Due to the stochastic process of rock displacement, the resistance in unit time n can be written as

$$R_n = r_0 - \sum_{h=1}^{n} D_h = r_0 - X_n, \quad n \in \mathbb{N}. \tag{2.13}$$

We consider *one* steel gate section and assume perfect inspection in the sense that the actual resistance can be determined without uncertainty.

Let the rock dumping be inspected at times $\{jk\Delta : j \in \mathbb{N}\}$ for $k \in \mathbb{N}$. Furthermore, inspection takes negligible time, does not degrade the rock dumping, and entails fixed costs c_I. We may regard the maintenance process as a renewal process, where renewals bring the rock dumping into the "as good as new state". Each renewal cycle ends either upon failure or at an inspection time $jk\Delta$ when the inspection reveals that a preventive repair should be carried out (for some $j \in \mathbb{N}$). A failure is defined as the event in which the resistance R drops below the failure level s: $R < s$. A preventive repair is defined as the event at which inspection reveals that the resistance has crossed the preventive repair level ρ while no failure has occurred: $s \leq R < \rho$, where $s < \rho < r_0$. A failure costs c_F Dutch guilders, while a preventive repair costs c_P Dutch guilders. Let the renewal times be conditionally independent random quantities having a discrete probability function $p_i(\theta, k)$, $i \in \mathbb{N}$, when the limiting average rate of rock displacement θ is given and the decision-maker chooses inspection decision k. The costs associated with a renewal at time $i\Delta$ are denoted by $c_i(\theta, k)$, $i \in \mathbb{N}$.

Since the planned lifetime of the barrier is very large, maintenance decisions can best be compared over an unbounded time-horizon. As Van Noortwijk & Peerbolte [29] have pointed out, there are basically three cost-based criteria that can serve as loss functions: (i) the expected average costs per unit time, (ii) the expected discounted costs over an unbounded time-horizon, and (iii) the expected equivalent average costs per unit time. These cost-based criteria can be obtained using the discrete renewal theorem (see e.g. Feller [13, Ch. 13] and Karlin & Taylor [15, Ch. 3]).

First, the expected *average costs per unit time* are determined by averaging the expected costs over an unbounded horizon:

$$L(\theta, k) = \lim_{n\to\infty} \frac{C(n,\theta,k)}{n} = \frac{\sum_{i=1}^{\infty} c_i(\theta,k)p_i(\theta,k)}{\sum_{i=1}^{\infty} i p_i(\theta,k)}, \tag{2.14}$$

where $C(n, \theta, k)$ are the expected costs in time-interval $(0, n\Delta]$. Eq. (2.14) is a well-known result from renewal reward theory (see e.g. Ross [23, Ch. 3]).

Second, the expected *discounted costs over an unbounded horizon* are determined by summing the expected discounted values of the costs over an unbounded horizon, where the discounted value of the costs c_n in unit time n is defined to be $\alpha^n c_n$ with discount factor $\alpha = [1 + (r/100)]^{-1}$ and discount rate $r\%$, where $r > 0$:

$$L_\alpha(\theta, k) = \lim_{n\to\infty} C_\alpha(n,\theta,k) = \frac{\sum_{i=1}^{\infty} \alpha^i c_i(\theta,k)p_i(\theta,k)}{1-\sum_{i=1}^{\infty} \alpha^i p_i(\theta,k)}, \tag{2.15}$$

where $C_\alpha(n, \theta, k)$ are the expected discounted costs in time-interval $(0, n\Delta]$.

Third, the expected *equivalent average costs per unit time* are determined by averaging the discounted costs over a "discounted" unbounded horizon (of length $1/(1-\alpha)$). In fact, the notion of equivalent average costs relates the notions of average costs and discounted costs in the sense that the equivalent average costs per unit time approach the average costs per unit time, as α tends to 1, from below:

$$\lim_{\alpha\uparrow 1}(1-\alpha)L_\alpha(\theta, k) = L(\theta, k). \tag{2.16}$$

The unit time for which the increments of rock displacement are distributed as mixtures of exponentials has been determined by specifying the conditional probability density function of the amount of rock displacement in a period of 50 years when the amount of erosion in a period of 100 years is given to be 70 displaced stones (using Eq. (2.4) shown in Fig. 6). When using the parameters of Table 2 (from Kok [17]) and applying numerical integration, the average costs per year and the equivalent average costs per year are represented by the curves in Fig. 7; the necessary expressions for deriving these costs can be found in Appendix 8. The optimal decision with respect to the criterion of average costs is $k^*\Delta = 10$ years, whereas the optimal decision with respect to the criterion of equivalent average costs is $k^*\Delta = 30$ years. Fortunately, the optimal inspection interval does not depend so much on the choice of the unit time (see Fig. 8). Recall that the maintenance costs are determined with respect to *one* of the 124 steel gate sections.

Table 2 The parameters of the maintenance model for the rock dumping.

parameter	description	value	dimension
Δ	unit time	5	year
Θ	average rate of rock displacement	$(0, \infty)$	stones/unit time
ν	shape parameter of $\text{Ig}(\theta\|\nu, \mu)$	12.2	
μ	scale parameter of $\text{Ig}(\theta\|\nu, \mu)$	39.2	
$E(\Theta)$	mean	3.5	stones/unit time
$\text{Var}(\Theta)$	variance	1.2	
r	discount rate per year	5	%
α	discount factor per unit time	0.7835	
c_I	costs of inspection	1,000	Dfl
c_P	costs of preventive repair	10,000	Dfl
c_F	costs of failure	120,000	Dfl
r_0	initial resistance	0	stones
ρ	preventive repair level	-50	stones
s	failure level	-70	stones
k	inspection-interval length	$\mathbb{N}$	unit time

5 CONCLUSIONS

In this chapter, we have presented two maintenance models that enable optimal inspection and repair decisions to be determined for two components of the Eastern-Scheldt barrier: the block mats and the rock dumping. Since there are only (subjective) probability distributions available on the average rates of scour erosion and rock displacement, we have based our models on these averages. In fact, we have shown that only with generalised gamma processes is one able to model stochastic deterioration processes with non-negative, exchangeable, real-valued increments when their average rates are uncertain. Two case studies have been carried out to show the usefulness of the maintenance models.

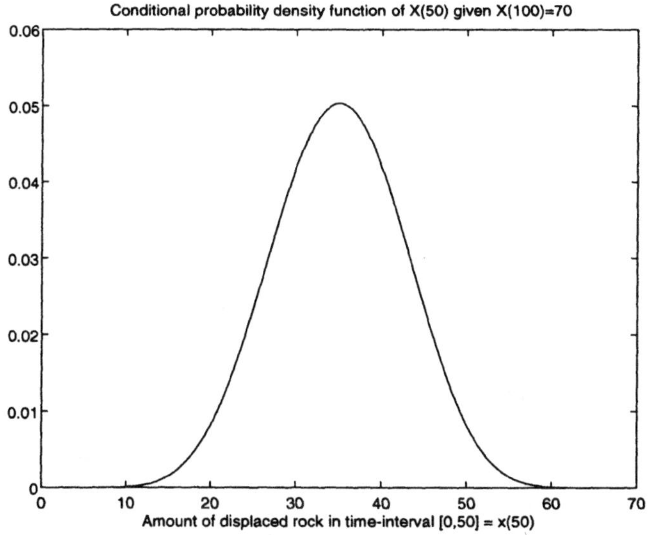

Figure 6 The conditional probability density function of the amount of rock displacement in a period of 50 years, $X(50)$, when the amount of rock displacement in a period of 100 years is $X(100) = 70$ stones.

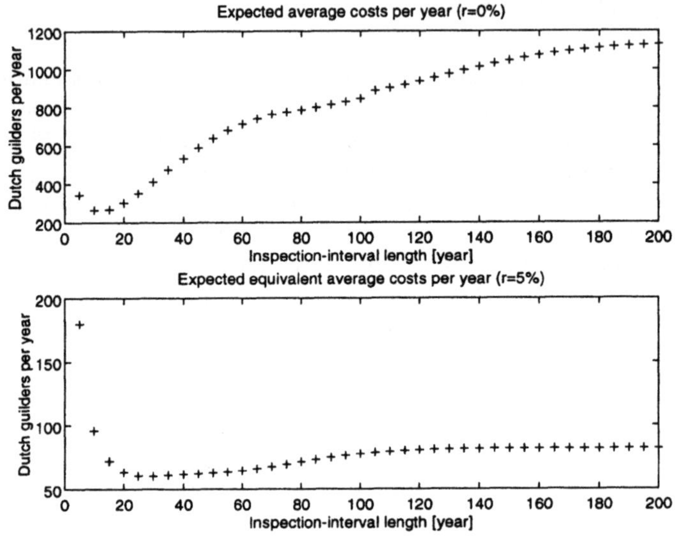

Figure 7 The expected average costs per year and the expected equivalent average costs per year for the rock dumping.

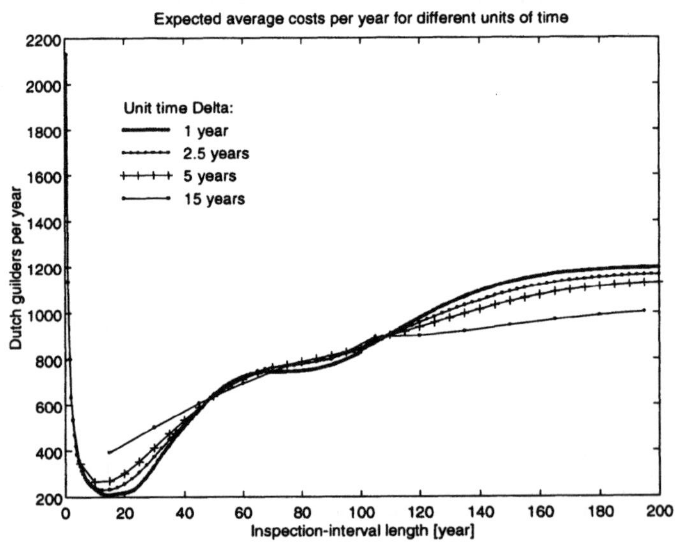

Figure 8 The expected average costs per year for different units of time Δ.

6 ACKNOWLEDGEMENTS

This work was partly supported by Delft Hydraulics, The Netherlands, under Project No. Q1210. The authors acknowledge helpful comments from Tim Bedford and they are grateful to Leo Klatter of the Civil Engineering Division of the Dutch Ministry of Transport, Public Works and Water Management for bringing the maintenance decision problem to their attention, providing the data, and giving fruitful comments.

7 APPENDIX: DEFINITIONS AND THEOREMS

Definition 1 (Gamma distribution.) *A random quantity X has a gamma distribution with shape parameter $a > 0$ and scale parameter $b > 0$ if its probability density function is given by:*

$$\text{Ga}(x \mid a, b) = [b^a / \Gamma(a)] \, x^{a-1} \exp\{-bx\} \, I_{(0,\infty)}(x).$$

Definition 2 (Inverted gamma distribution.) *A random quantity Y has an inverted gamma distribution with shape parameter $a > 0$ and scale parameter $b > 0$ if $X = Y^{-1} \sim \text{Ga}(a, b)$. Hence, the probability density function of Y is:*

$$\text{Ig}(y \mid a, b) = [b^a / \Gamma(a)] \, y^{-(a+1)} \exp\{-b/y\} \, I_{(0,\infty)}(y).$$

Definition 3 (Beta distribution.) *A random quantity X has a beta distribution with parameters $a, b > 0$ if its probability density function is given by:*

$$\text{Be}(x|a, b) = \frac{\Gamma(a+b)}{\Gamma(a)\Gamma(b)} \, x^{a-1}(1-x)^{b-1} \, I_{[0,1]}(x).$$

Definition 4 (Negative multinomial distribution.) *An n-dimensional random vector $\mathbf{Y}$, where $\mathbf{Y} = (Y_1, \ldots, Y_n)$, has a negative multinomial distribution with parameters $\mathbf{p} = (p_1, \ldots, p_n)$ and ν if $\mathbf{Y}$ has a probability function given by:*

$$\text{Nm}(\mathbf{y}|\mathbf{p}, \nu) = \frac{\Gamma\left(\nu + \sum_{i=1}^{n} y_i - n\right)}{\Gamma(\nu) \prod_{i=1}^{n} \Gamma(y_i)} \left(1 - \sum_{i=1}^{n} p_i\right)^{\nu} \prod_{i=1}^{n} p_i^{y_i - 1},$$

$y_i = 1, 2, \ldots$ *for* $i = 1, \ldots, n$, $\nu > 0$, $p_i > 0$ *for all* i, *and* $\sum_{i=1}^{n} p_i < 1$.

Theorem 1 *Let Λ and Θ be independent inverted gamma distributed random quantities such that $\Lambda \sim \mathrm{Ig}(\alpha, \beta)$ and $\Theta \sim \mathrm{Ig}(\nu, \mu)$ for $\alpha, \nu > 2$ and $\beta, \mu > 0$. Furthermore, let $(Y_1, Y_2) \sim \mathrm{Nm}(y/(\mu + 2y), y/(\mu + 2y), \nu)$ for $y > 0$ (implying the marginal distribution of Y_1 to have a negative binomial distribution: $Y_1 \sim \mathrm{Nm}(y/(\mu + y), \nu)$). Then,*

$$\begin{aligned} \xi_{k1} - \xi_{k2} &< 1 - E\left(\exp\left\{-\frac{\Delta}{\Lambda}\sum_{i=1}^{k}\sum_{h=1}^{k-i+1}\frac{1}{(h-1)!}\left[\frac{y}{\Theta}\right]^{h-1}\exp\left\{-\frac{y}{\Theta}\right\}\right\}\right) \\ &< \xi_{k1}, \end{aligned} \tag{2.17}$$

where

$$\xi_{km} = \frac{1}{m}E\left(\left[\frac{\Delta}{\Lambda}\right]^m\right)\left[\sum_{i=1}^{k}\sum_{h=1}^{k-i+1}\left(\Pr\{Y_1 = h, Y_m = h\}\right)^{\frac{1}{m}}\right]^m$$

for $m = 1, 2$, $k \in \mathbb{N}$ and $\Delta > 0$.

Proof:

The upper bound in Eq. (2.17), ξ_{k1}, can be derived by using the inequality $1 - e^{-x} < x$ for $x > 0$ and the gamma integral.

The lower bound in Eq. (2.17), $\xi_{k1} - \xi_{k2}$, can be derived by using the inequality $1 - e^{-x} > x - \frac{1}{2}x^2$ for $x > 0$, Minkowski's inequality for integrals, and the gamma integral. □

Theorem 2 *Let $\{Y_i : i \in \mathbb{N}\}$ be an infinitely exchangeable sequence of positive real-valued random quantities such that*

$$\Pr\{(Y_1, \ldots, Y_n) \in A\} = \Pr\{(Y_1, \ldots, Y_n) \in A + \mathbf{x}\}$$

for all $n \in \mathbb{N}$ and any Borel set $A \in \mathbb{R}_+^n$ with $\mathbf{x} \in \mathbb{R}^n$ satisfying $\sum_{i=1}^n x_i = 0$ and $A + \mathbf{x} \in \mathbb{R}_+^n$. Then, the joint probability density function of $(Y_1, \ldots, Y_n)$ is a scale mixture of exponentials for all $n \in \mathbb{N}$. If in addition, $E(Y_2|Y_1 = y_1) = ay_1 + b$ for some constants $a, b > 0$, then the mixing measure is a gamma distribution.

Proof:

See Diaconis & Freedman [10] and Diaconis & Ylvisaker [12]. □

Theorem 3 *Let $\{Y_i : i \in \mathbb{N}\}$ be an infinitely exchangeable sequence of real-valued random quantities such that, for all $n \geq 2$ and $k < n$,*

$$p(y_1, \ldots, y_k | \textstyle\sum_{i=1}^n y_i = t) = \frac{\left[\prod_{i=1}^k h(y_i)\right] h^{(n-k)}\left(t - \sum_{i=1}^k y_i\right)}{h^{(n)}(t)}, \tag{2.18}$$

where $h^{(n)}$ is the n-fold convolution of the (positive and continuous) function h with itself and

$$c(\theta) = \int h(y) \exp\{\theta y\}\, dy < \infty \quad \textit{for } \theta \in \mathbb{R}.$$

Then, there exists a probability distribution P such that

$$p(y_1, \ldots, y_n) = \int \prod_{i=1}^n [h(y_i) \exp\{\theta y_i\}/c(\theta)]\, dP(\theta).$$

Proof:

See Diaconis & Freedman [9]. □

Theorem 4 *Let $\{X(t) : t \geq 0\}$ be a non-decreasing continuous-time stochastic process with $X(0) = 0$, with probability one, such that for every $\tau > 0$ the infinite sequence of non-negative real-valued increments $D_i(\tau) = X(i\tau) - X([i-1]\tau)$, $i \in \mathbb{N}$, is exchangeable. Moreover, for every $\tau > 0$, and all $n \geq 2$ and $k < n$, the joint conditional probability density function of the increments $D_1(\tau), \ldots, D_k(\tau)$, when $X(n\tau) = x$ is given, can be represented by*

$$p_{D_1(\tau),\ldots,D_k(\tau)|X(n\tau)}(\delta_1, \ldots, \delta_k | x) = $$

$$= \frac{\left[\prod_{i=1}^k h(\delta_i, \tau)\right] h^{(n-k)}\left(x - \sum_{i=1}^k \delta_i, \tau\right)}{h^{(n)}(x, \tau)}, \tag{2.19}$$

where $h(x, \tau)$ is differentiable and non-negative, $h^{(n)}(x, \tau)$ is the n-fold convolution in x of $h(x, \tau)$ with itself, and $c(\theta)$ is defined by

$$\int_0^\infty h(x, \tau) c(\theta) \exp\{-x/\theta\}\, dx = \int_0^\infty l(x|\theta)\, dx = 1 \tag{2.20}$$

for $\theta \in (0, \infty)$. In addition, let the likelihood function $l(x|\theta)$ be a scale density:

$$l(x|\theta) = f(x/\theta)/\theta \quad \textit{for } x, \theta \in (0, \infty). \tag{2.21}$$

Then there exists a constant $a > 0$ such that the joint probability density function of the increments $D_1(\tau), \ldots, D_n(\tau)$ is given by Eq. (2.5).

Proof:

By Diaconis & Freedman [9] and Küchler & Lauritzen [18] (see also Theorem 3), there exists a probability distribution $P_{\Theta(\tau)}$ such that

$$p_{D_1(\tau),\dots,D_n(\tau)}(\delta_1,\dots,\delta_n) = \int_0^\infty \prod_{i=1}^n h(\delta_i,\tau)c(\theta)\exp\{-\delta_i/\theta\}\,dP_{\Theta(\tau)}(\theta)$$

for every $\tau > 0$ and all $n \in \mathbb{N}$. Since the likelihood function $l(x|\theta)$ is a scale density, the function $h(x,\tau)$ satisfies the functional equation

$$l(x|\theta) = h(x,\tau)c(\theta)\exp\{-x/\theta\} = f(x/\theta)/\theta$$

or, with $g(x/\theta) = \exp\{x/\theta\}f(x/\theta)$, $\phi_1(x) = h(x,\tau)$, and $\phi_2(\theta) = \theta c(\theta)$:

$$g(x/\theta) = \phi_1(x)\phi_2(\theta).$$

This functional equation can be recognised as an extension of one of the four well-known Cauchy equations in which $g(x) = \phi_1(x) = \phi_2(1/x)$ (see Huzurbazar [14, p. 204]). Its general solution is $g(x) = c_1 x^{c_2}$, where c_1 and c_2 are arbitrary constants. Hence, using Eq. (2.20), the functions $h(x,\tau)$ and $c(\theta)$ have the form

$$h(x,\tau) = x^{\alpha(\tau)}/\Gamma(\alpha(\tau)+1), \qquad c(\theta) = \theta^{-\alpha(\tau)-1},$$

respectively, where $\alpha(\tau) > -1$ is a differentiable function. In turn, $\alpha(\tau)$ satisfies another Cauchy functional equation: $\alpha(n\tau) = n\alpha(\tau)$ for all $\tau > 0$ and $n \in \mathbb{N}$. This functional equation is generated by Eq. (2.19), i.e.

$$p_{D_1(n\tau)|X(2n\tau)}(\delta|\,x) = p_{D_1(\tau)+\dots+D_n(\tau)|X(2n\tau)}(\delta|\,x), \tag{2.22}$$

when dividing both sides of Eq. (2.22) by $\delta^{n\alpha(\tau)}$ and letting δ approach zero from the right. The general solution is $\alpha(\tau) = a\tau + b$ for some constants $a > 0$ and $b \geq -1$. Since $X(0) = 0$, with probability one, we have $b = -1$.
Eq. (2.5) follows by replacing θ with $\theta/(a\tau)$, which proves the theorem. □

8 APPENDIX: THE EXPECTED MAINTENANCE COSTS

In order to compare maintenance decisions over an unbounded horizon for the rock dumping of the Eastern-Scheldt barrier, we need to determine two cost-based criteria: (i) the expected average costs per unit time, Eq. (2.14), and (ii) the expected discounted costs over an unbounded horizon, Eq. (2.15). For this purpose, expressions have been derived for the expected cycle costs, the expected cycle length, the expected discounted cycle costs, and the expected "discounted cycle length". They can be determined using the failure model of Van Noortwijk, Cooke & Kok [28]. For notational convenience, let $x = r_0 - \rho$, $y = r_0 - s$, and

$$\psi_{j,n} = \sum_{i=j+1}^{n} \binom{n-1}{i-1} \left[1 - \frac{x}{y}\right]^{n-i} \left[\frac{x}{y}\right]^{i-1}$$

for $j < n$. Recall that inspections are scheduled at times $\{jk\Delta : j \in \mathbb{N}\}$ with inspection interval $k \in \mathbb{N}$. The costs of inspection are c_I, the costs of failure are c_F, and the costs of preventive repair are c_P.

The expected cycle costs.
The expected cycle costs can be written as the sum of the expected costs due to inspection, preventive repair, and failure (using Eq. (2.11)):

$$\sum_{i=1}^{\infty} c_i(\theta, k) p_i(\theta, k) =$$

$$= \sum_{j=1}^{\infty} [jc_I + c_P] \Pr\left\{ R_{(j-1)k} \geq \rho, s \leq R_{jk} < \rho \,\middle|\, \theta \right\} +$$

$$\sum_{j=1}^{\infty} [(j-1)c_I + c_F] \Pr\left\{ R_{(j-1)k} \geq \rho, R_{jk} < s \,\middle|\, \theta \right\}$$

$$= \sum_{j=1}^{\infty} \sum_{n=(j-1)k+1}^{jk} \left\{ [jc_I + c_P]\, q_n(\theta, x) + [c_F - c_P - c_I]\, \psi_{(j-1)k,n} q_n(\theta, y) \right\}.$$

The expected cycle length.
Similarly, the expected cycle length can be written as

$$
\begin{aligned}
&\sum_{i=1}^{\infty} i p_i(\theta, k) = \\
&\quad = \sum_{j=1}^{\infty} jk \Pr\left\{ R_{(j-1)k} \geq \rho, s \leq R_{jk} < \rho \,\middle|\, \theta \right\} + \\
&\quad\quad \sum_{j=1}^{\infty} \sum_{n=(j-1)k+1}^{jk} n \Pr\left\{ R_{(j-1)k} \geq \rho, R_{n-1} \geq s, R_n < s \,\middle|\, \theta \right\} \\
&\quad = \sum_{j=1}^{\infty} \sum_{n=(j-1)k+1}^{jk} \left\{ jk q_n(\theta, x) + (n - jk)\psi_{(j-1)k,n} q_n(\theta, y) \right\}.
\end{aligned}
$$

The expected discounted cycle costs.
The expected discounted cycle costs can be written as the discounted value of the expected costs due to inspection, preventive repair, and failure:

$$
\begin{aligned}
&\sum_{i=1}^{\infty} \alpha^i c_i(\theta, k) p_i(\theta, k) = \\
&= \sum_{j=1}^{\infty} \left[\left(\sum_{h=1}^{j} \alpha^{hk} \right) c_I + \alpha^{jk} c_P \right] \Pr\left\{ R_{(j-1)k} \geq \rho, s \leq R_{jk} < \rho \,\middle|\, \theta \right\} + \\
&\quad \sum_{j=1}^{\infty} \sum_{n=(j-1)k+1}^{jk} \left[\left(\sum_{h=1}^{j-1} \alpha^{hk} \right) c_I + \alpha^n c_F \right] \Pr\left\{ R_{(j-1)k} \geq \rho, R_{n-1} \geq s, R_n < s \,\middle|\, \theta \right\} \\
&= \sum_{j=1}^{\infty} \sum_{n=(j-1)k+1}^{jk} \left[\left(\frac{1 - \alpha^{jk}}{1 - \alpha^k} \right) \alpha^k c_I + \alpha^{jk} c_P \right] q_n(\theta, x) + \\
&\quad \sum_{j=1}^{\infty} \sum_{n=(j-1)k+1}^{jk} \left(\alpha^n c_F - \alpha^{jk} [c_I + c_P] \right) \psi_{(j-1)k,n} q_n(\theta, y).
\end{aligned}
$$

The expected "discounted cycle length".
Similarly, the expected "discounted cycle length" can be written as:

$$\sum_{i=1}^{\infty} \alpha^i p_i(\theta, k) =$$

$$= \sum_{j=1}^{\infty} \alpha^{jk} \Pr\left\{ R_{(j-1)k} \geq \rho, s \leq R_{jk} < \rho \,\middle|\, \theta \right\} +$$

$$\sum_{j=1}^{\infty} \sum_{n=(j-1)k+1}^{jk} \alpha^n \Pr\left\{ R_{(j-1)k} \geq \rho, R_{n-1} \geq s, R_n < s \,\middle|\, \theta \right\}$$

$$= \sum_{j=1}^{\infty} \sum_{n=(j-1)k+1}^{jk} \left\{ \alpha^{jk} q_n(\theta, x) + \left(\alpha^n - \alpha^{jk}\right) \psi_{(j-1)k,n} q_n(\theta, y) \right\}.$$

REFERENCES

[1] Richard E. Barlow and Max B. Mendel. De Finetti-type representations for life distributions. *Journal of the American Statistical Association*, 87(420):1116–1122, December 1992.

[2] Richard E. Barlow and Frank Proschan. *Mathematical Theory of Reliability*. John Wiley & Sons, 1965.

[3] José M. Bernardo and Adrian F.M. Smith. *Bayesian Theory*. John Wiley & Sons, Inc., 1994.

[4] Danny I. Cho and Mahmut Parlar. A survey of maintenance models for multi-unit systems. *European Journal of Operational Research*, 51:1–23, 1991.

[5] Yuan Shih Chow and Henry Teicher. *Probability Theory: Independence, Interchangeability, Martingales;* Second Edition. Springer-Verlag, 1988.

[6] J.J. de Jonge, M. Kok, and J.M. van Noortwijk. Onderhoud van de natte werken van waterbouwkundige constructies [Maintenance of the wet works of hydraulic structures]. Technical Report Q1297, Delft Hydraulics, The Netherlands, November 1991.

[7] Morris H. DeGroot. *Optimal Statistical Decisions*. McGraw-Hill Book Company, 1970.

[8] Rommert Dekker. Applications of maintenance optimization models: a review and analysis. *Reliability Engineering and System Safety*, 51:229–240, 1996.

[9] P. Diaconis and D.A. Freedman. Cauchy's equation and de Finetti's theorem. *Scandinavian Journal of Statistics*, 17:235–250, 1990.

[10] Persi Diaconis and David Freedman. A dozen de Finetti-style results in search of a theory. *Annales de l'Institute Henri Poincaré*, 23(2):397–423, 1987.

[11] Persi Diaconis and Donald Ylvisaker. Conjugate priors for exponential families. *The Annals of Statistics*, 7(2):269–281, 1979.

[12] Persi Diaconis and Donald Ylvisaker. Quantifying prior opinion. In J.M. Bernardo, M.H. DeGroot, D.V. Lindley, and A.F.M. Smith, editors, *Bayesian Statistics 2*, pages 133–156. Elsevier Science Publishers B.V., 1985.

[13] William Feller. *An Introduction to Probability Theory and its Applications; Volume 1.* John Wiley & Sons, 1950.

[14] Vasant S. Huzurbazar. *Sufficient Statistics: Selected Contributions.* Marcel Dekker, Inc., 1976.

[15] Samuel Karlin and Howard M. Taylor. *A First Course in Stochastic Processes;* Second Edition. Academic Press, Inc., 1975.

[16] M. Kok. Stormvloedkering Oosterschelde; optimaliseren van inspectie en onderhoud natte werken [Eastern-Scheldt storm-surge barrier; optimising inspection and maintenance of the wet works]. Technical Report Q606, Delft Hydraulics, The Netherlands, January 1988.

[17] M. Kok. Onderhoud; methoden voor rationeel onderhoud van civiele constructies [Maintenance; methods for rational maintenance of civil structures]. Technical Report Q606, Delft Hydraulics, The Netherlands, September 1990.

[18] Uwe Küchler and Steffen L. Lauritzen. Exponential families, extreme point models and minimal space-time invariant functions for stochastic processes with stationary and independent increments. *Scandinavian Journal of Statistics*, 16:237–261, 1989.

[19] J.J. McCall. Maintenance policies for stochastically failing equipment: a survey. *Management Science*, 11:493–524, 1965.

[20] Jolanta K. Misiewicz and Roger M. Cooke. Isotropy and stochastic rescaling. *Sankhyã: The Indian Journal of Statistics. Series A: Probability and Mathematical Statistics.* (to appear), 1996.

[21] William P. Pierskalla and John A. Voelker. A survey of maintenance models: the control and surveillance of deteriorating systems. *Naval Research Logistics Quarterly*, 23:353–388, 1976.

[22] Rijkswaterstaat. *Design plan Oosterschelde storm-surge barrier; overall design and design philosophy.* Ministry of Transport, Public Works and Water Management, Delft, The Netherlands, 1994.

[23] Sheldon M. Ross. *Applied probability models with optimization applications.* Holden-Day, 1970.

[24] Y. S. Sherif and M.L. Smith. Optimal maintenance models for systems subject to failure - a review. *Naval Research Logistics Quarterly*, 28:47–74, 1981.

[25] Yosef S. Sherif. Reliability analysis: optimal inspection and maintenance schedules of failing systems. *Microelectron. Reliability*, 22(1):59–115, 1982.

[26] Lennaert J.P. Speijker, Jan M. van Noortwijk, Roger M. Cooke, and Matthijs Kok. Optimal maintenance decisions for dykes. Technical Report 96-39, Faculty of Mathematics and Computer Science, Delft University of Technology, The Netherlands, 1996.

[27] Ciriaco Valdez-Flores and Richard M. Feldman. A survey of preventive maintenance models for stochastically deteriorating single-unit systems. *Naval Research Logistics*, 36:419–446, 1989.

[28] Jan M. van Noortwijk, Roger M. Cooke, and Matthijs Kok. A Bayesian failure model based on isotropic deterioration. *European Journal of Operational Research*, 82(2):270–282, April 1995.

[29] Jan M. van Noortwijk and E. Bart Peerbolte. Optimal sand nourishment decisions. Publication 493, Delft Hydraulics, The Netherlands, November 1995.

[30] Jan M. van Noortwijk and Pieter H.A.J.M. van Gelder. Optimal maintenance decisions for berm breakwaters. Technical Report 96-28, Faculty of Mathematics and Computer Science, Delft University of Technology, The Netherlands, 1996.

[31] Ian Watson and Charles W. Finkl Jnr. Simplified technical summary of the complete Delta Works, including the Eastern Scheldt. *Journal of Coastal Research, Special Issue No. 10*, 1992.

3

REVIEW PROBLEM OWNER PERSPECTIVE

Leo Klatter

Civil Engineering Division, Rijkswaterstaat, Utrecht, The Netherlands

1 INTRODUCTION

In this review of the paper "Optimal Maintenance Decisions for the Sea-Bed Protection of the Eastern-Scheldt Barrier," a profile of the problem owner and some general characteristics of the sea-bed protection will first be given. After this, the two examples presented in the paper will be reviewed.

2 PROFILE PROBLEM OWNER

The "problem owner" is the regional division of Rijkswaterstaat in the province of Zeeland. This division is responsible for the operational maintenance of the Eastern-Scheldt Barrier. The maintenance staff has a large practical experience with structures like the sea-bed protection. Such structures are frequently used in the Netherlands. The Civil Engineering Division advises the regional division on special design aspects of the sea-bed protection. For these aspects a committee has been installed that meets every three months. There is evidently a large gap in mathematical knowledge between the maintenance staff and mathematicians who are modelling the maintenance problems. This gap can be reduced, using the more scientific based design knowledge of the advisor of the maintenance staff. The most difficult task is to establish an effective communication between mathematicians and the maintenance staff, to provide the mathematicians with the data needed to model the problem and to establish confid^nce in the results of the model by the maintenance staff.

R. Cooke et al. (eds.), Engineering Probabilistic Design and Maintenance for Flood Protection, 57–59.

3 GENERAL COMMENTS ON THE MODEL

The characteristics of the sea-bed protection are:

- a long design lifetime (50 - 200 years)
- large uncertainties in the actual strength of the structure and the expected loads on the structure
- unique, tailor-made structures
- the condition of the structure cannot be determined accurately by inspections

These characteristics call for a model of the deterioration proces that can cope with large uncertainties and that shows a transparent response to maintenance measures, so one can understand the results. For this type of problems a sensitivity analysis is often even more important, than the exact result of the model. The deterioration proces is often too complex to be described entirely by the model. The model described in the paper solves only parts of the maintenance problem. This approach is, given the nature of the problem, a preferable approach.

4 MAINTENANCE OF THE BLOCK MATS

The actual maintenance problem that had to be solved was: *To what extent can the inspections be reduced without endangering the safety of the barrier itself?*

The part of the sea-bed protection involved is at some distance from the barrier axis, where a certain rate of scour can be accepted. The safety of the barrier determines the maximum acceptable depth of the scour hole. The model had to be able to compute the probability that this maximum scour depth would be exceeded. From this, the minimum inspection frequency can be determined. A more frequent inspection can be economical comparing the expected repair costs and the inspection costs. The results of the model can be used to investigate the sensitivity of the inspection frequency to the occurrence rate of scour holes quite well. Also the sensitivity of the inspection frequency to the uncertainties in strength, load and inspection accuracy could be investigated effectively. The scour rate is linearized in the model. Since we were only interested in a limited range of scour hole depths, in practice between 5 and 15 meters, this is acceptable.

5 MAINTENANCE OF THE ROCK DUMPING

The maintenance problem to be solved was: *What is the inspection criterion for over-loading of the rock structure between the rubble sill of the barrier and the sea-bed protection?*

From model tests it is known that above a certain level of loading (as a measure for the hydraulic loading the water level difference over the barrier is used) a number of stones will be displaced each time an over-load occurs. The cumulative number of displaced rock is not allowed to exceed a certain limit. An inspection has to be planned before this limit is reached. The criterion for this inspection had to be determined. Translated into statistical properties, using the probability density function of these loads, an inspection load (expressed as a water-level difference) corresponds to an inspection interval. With the model an economical inspection criterion could be determined. All costs presented in Table 2 are based on a small area of the structure. The costs of failure in the table are defined as the probability of failure of the barrier, with a damage level of -70 stones, multiplied by the costs of failure of a part of the barrier.

6 CONCLUSIONS

An effective communication between the mathematicians and the maintenance staff is necessary for a succesful use of a model as described in the paper to obtain input data for the model and the acceptance of the results. A strong effort is still necessary on this subject.

The model described in the paper was able to cope with the large uncertainties related to modelling deterioration of sea-bed protections and gave realistic solutions to the problems that were modelled.

4

REVIEW ENGINEERING PERSPECTIVE

John Shortle

Department of Industrial Engineering and Operations Research, University of California, Berkeley, California 94720

ABSTRACT

The main paper gives a model for the initiation and development of scour holes in the Eastern-Scheldt Barrier. The *initiation* of scour holes is modelled as a Poisson process. The *erosion* of these holes, once started, is modelled as a gamma process. This review addresses how well the Poisson and gamma processes model scour erosion at the Eastern-Scheldt Barrier.

1 SCOUR HOLE INITIATION

The initiation of scour holes is modelled as a Poisson process. This section gives suggestions for non-stationary extensions of the model.

The Poisson process has stationary and independent increments. The stationarity assumption implies a *constant* rate of occurrence of scour holes. In reality, if the protective mat decays, this rate increases over time. For instance, a large storm may damage the mat, making it more susceptible to scour holes in the future. The constant tidal flow over the mat may also weaken it over time. Therefore, it is expected that the number of scour holes per year increases.

So far (1995), no scour holes have been observed. From this and the prior, the model gives that the mean interoccurrence time of scour holes is 3.7 years. Now, consider the following scenario (Figure 1): In 1996, five scour holes are observed, and in 1997, seven are observed. How many scour holes appear after that? The model gives that the new mean interoccurrence time of scour holes

R. Cooke et al. (eds.), Engineering Probabilistic Design and Maintenance for Flood Protection, 61–66.

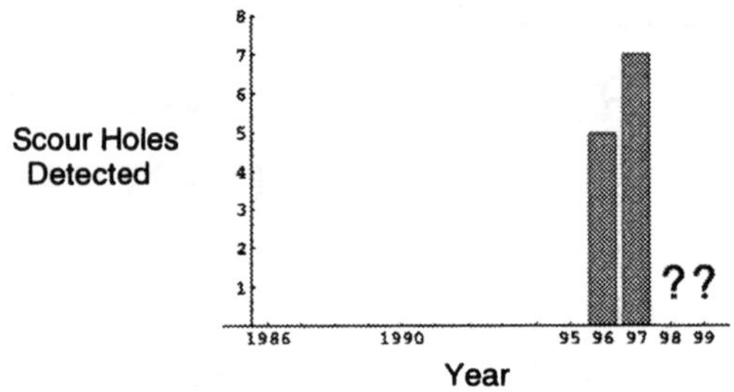

Figure 1 No scour holes have been observed from 1986-1995. What does the model predict if five holes are observed in 1996 and seven in 1997?

is 1.30 years - less than one scour hole per year! In reality, seven or eight holes may appear in 1998. It is questionable whether or not the first ten years of data are even relevant any more.

Thus, the situation is fundamentally non-stationary. Yet, it is difficult to estimate the rate $\lambda(t)$ of a non-stationary Poisson process. As pointed out in the main paper, it is difficult enough to estimate a *constant* rate λ.

Some work has been done in Bayesian estimation of this non-stationary rate. A common approach is to assume a parametric form for the rate function $\lambda(t)$ and to update the parameters using Bayes' rule. For instance, Clevenson [2] assumes a linear form. Calabria [1] assumes a power law form:

$$\lambda(t) = (\beta/\alpha)(t/\alpha)^{\beta-1}.$$

Non-parametric models also exists. Lo [5], for instance, shows how a prior gamma distribution can be used to generate a random rate function $\lambda(t)$. The posterior $\lambda(t)$ is similarly a random function generated by an updated gamma distribution. Silver [6] gives another kind of model in which $\lambda(t)$ is assumed constant over short intervals. Bayes' rule is used to determine if $\lambda(t)$ changes from one interval to the next. This model works well for processes with sudden irreversible rate changes (for instance, large storms suddenly weakening the mat).

2 SCOUR EROSION

The erosion process of scour holes, once begun, is modelled as a gamma process. This section summarizes empirical results from research on scour erosion. Sev-

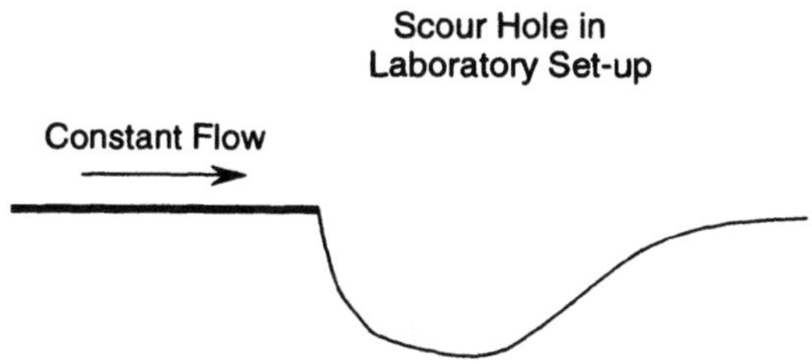

Figure 2 A typical setup for the empirical study of scour hole growth.

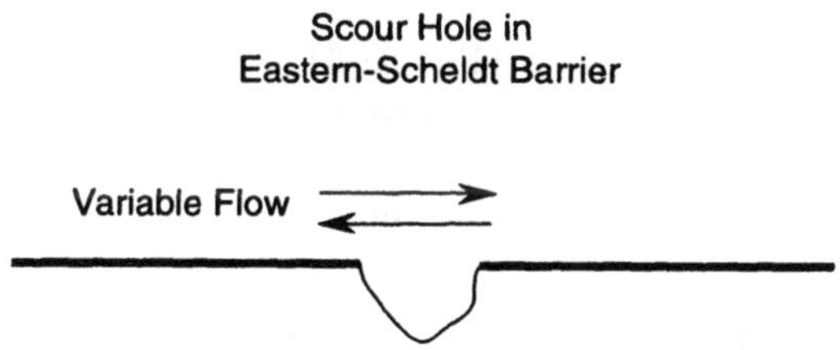

Figure 3 Scour erosion at the Eastern-Scheldt Barrier.

eral questions are posed in relation to these results. Properties of the gamma process are examined in the context of scour erosion.

Two surveys of empirical research in scour erosion are presented in Hoffmans [4] and Einstein [3]. Scour erosion occurs when there is no sedimentary supply to an area - for instance, when a dam obstructs downstream transport of sand. This is typically set up in the laboratory as follows: A mat is placed on a bed of sand, and a constant stream of water is sent over the mat (Figure 2). Scouring begins at the edge of the mat. The depth of the scour hole is measured as a function of time. Studies by Delft Hydraulics, for instance, give the following empirical relation (Hoffmans [4], p. 148):

$$X(t) \propto t^{\alpha},$$

where $X(t)$ is the maximum scour depth at time t, and α is a constant *less than* 1. The value of α depends on sedimentary characteristics and other parameters. Thus, the rate of erosion *decreases* over time. Other empirical studies show similar results (for example, Einstein [3], p. 5-19).

The main paper assumes a *constant* rate of scour erosion. This is not necessarily inconsistent with the laboratory results, since the situation at the Eastern-

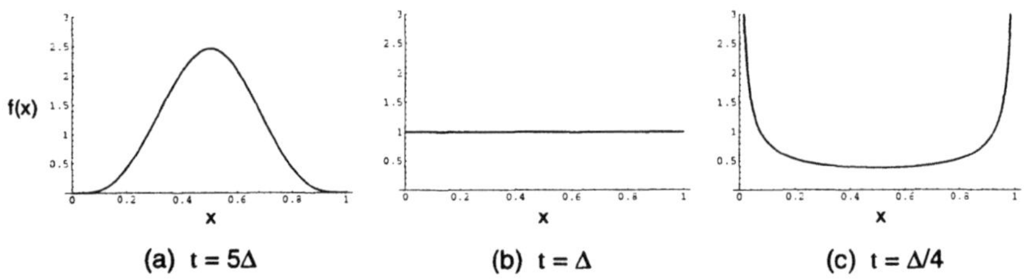

Figure 4 Plots of $f_{(X(t/2)|X(t)=1)}(x)$ for different values of t.

Scheldt is different in several ways (Figure 3). First of all, the flow over the mat changes in magnitude and direction due to the change of tides. Secondly, erosion occurs where the protective mat is damaged, rather than where the mat ends. Some questions to consider about the Eastern-Scheldt barrier are:

1. Can scour holes fill back up?
2. Is there a maximum depth to the scour holes? For instance, if a 1 meter hole develops *in the mat*, can the scour hole get any deeper than, say, 2 meters?
3. Or, does damage to the mats also increase with time?
4. And if so, which is more critical: the rate of damage to the mats or the rate at which sand is carried away?

For instance, if the holes in the mat remain fixed, then the empirical results suggest that the scour rate decreases over time. In this case, the model overestimates the damage actually done. On the other hand, if the holes in the mat increase, then a constant scour rate may be a good assumption.

Another property of the gamma process is that the increments of erosion are l_1-isotropic, for some time interval Δ. That is, the joint distribution of the increments $f(D_1(\Delta), D_2(\Delta), \cdots, D_n(\Delta))$ depends only on $\sum_{i=1}^{n} D_i(\Delta)$. This means, for instance, that the following two events are *equally likely*: (a) observing 1 meter of erosion for 100 consecutive time units Δ, and (b) observing 100 meters of erosion in 1 time unit Δ and no erosion in the other 99. This is a questionable assumption, but a closer look reveals otherwise.

Suppose one meter of damage has been observed by time t. The distribution of damage by time $t/2$ is:

$$f_{(X(t/2)|X(t)=1)}(x) \propto x^{\frac{t}{\Delta}-1}(1-x)^{\frac{t}{\Delta}-1}. \tag{4.1}$$

This is the beta distribution. We check this result against our intuition. If t is large, say one year, then we expect the damage after a half year to be close to half of the total damage. Such a density might look something like Figure 4a.

On the other hand, if t is small, say one hour, we expect the total damage to occur either in the first half hour or in the second. That is, we believe that damage occurs mostly in chunks on this time scale. Such a density might look something like Figure 4c.

It is then reasonable to assume that there is some intermediate value of t for which the density is flat (Figure 4b). In this unit of time, the increments are l_1-isotropic. Thus, the gamma process is consistent with our intuition in this sense. Considering the previous empirical results, a generalization would be to model the erosion process as a gamma process with decreasing rate.

REFERENCES

[1] R. Calabria, M. Guuida, and G. Pulcini. Bayes estimation of prediction intervals for a power law process. *Communications in Statistics. Theory and Methods*, 19(8):3023–3035, 1990.

[2] M. L. Clevenson and J. W. Zidek. Bayes linear estimators of the intensity function of the nonstationary Poisson process. *Journal of American Statistical Association*, 72:112–120, 1977.

[3] H. A. Einstein and R. L. Wiegel. A literature review on erosion and deposition of sediment near structures in the ocean. Technical Report 21-6, University of California, Berkeley, 1970. Hydraulic Engineering Laboratory, College of Engineering.

[4] G. J. C. M. Hoffmans. Two-dimensional mathematical modelling of local-scour holes. Technical Report 92-7, Delft University of Technology, 1992. Series title: Communications on Hydraulic and Geotechnical Engineering.

[5] Albert Y. Lo. Bayesian nonparametric statistical inference for Poisson point processes. *Zeitschrift für Wahrscheinlichkeitstheorie und verwandte Gebiete*, 59:55–66, 1982.

[6] Edward A. Silver. Bayesian modelling of a non-stationary Poisson process. *Canadian Journal of Operational Research and Information Processing*, 9:32–45, 1971.

5

GAMMA PROCESSES AND THEIR GENERALIZATIONS: AN OVERVIEW

Nozer Singpurwalla

Department of Engineering and Operations Research, The George Washington University, Washington, DC 20052

ABSTRACT

The deterioration and erosion of many civil engineering structures and other structural components, such as the skin of aircraft, have been described by gamma processes and their generalizations. The objective of this article is to overview such processes and to present their salient features such as decomposition and representation. The overview is not necessarily complete nor is it self-contained. The article begins with a motivating example from a problem in hydrological engineering.

Keywords. Deterioration, erosion, infinite divisibility, jump processes, Lévy decomposition, representation theorems.

1 INTRODUCTION

This overview has been prompted by a discussion of a paper dealing with the scheduling of optimal inspection of the sea-bed protection of the Dutch Eastern-Scheldt barrier. The sea-bed protection is subject to deterioration due to the presence of scour holes. The deterioration has been described by a gamma process. Broadly speaking, a motivation for the consideration of gamma processes for describing the deterioration due to scour holes is the following.

R. Cooke et al. (eds.), Engineering Probabilistic Design and Maintenance for Flood Protection, 67–75.

1.1 Motivation: The Hydrological Problem

A certain sea-bed protection, of critical importance, experiences the presence of scour holes which occur over time. Let $T_1, T_2, \cdots$, be the times between occurrence of the scour holes. The T_i's are assumed to be infinite in number and are also judged to be exchangeable. Exchangeability induces a mild form of dependence between the T_i's. Each scour hole triggers a deterioration mechanism in the sea-bed protection. The deterioration mechanism weakens the protection over time, and eventually makes it unsuitable for use. Maintenance actions could arrest the growth of deterioration, also known as "erosion" but this is not of concern to us here. Rather, what is of interest is the scheduling of inspections in an optimal manner so that a potential catastrophe can be avoided. Inspection and maintenance are difficult and time consuming operations; they are also extremely cumbersome involving a shutdown of the hydrological system. Figure 1 depicts the scenario described above; here $S_n = \sum_{i=1}^{n} T_i$.

Upon inspection one observes two quantities: the number of scour holes and the deterioration of each. For example, in Figure 1, at Inspect (3) one observes three scour holes, having a deteriorations X_1, X_2, and X_3. Clearly, in order to obtain an optimal inspection schedule, we need to know, besides the necessary costs of inspections and losses due to catastrophes, the stochastic behavior of the number of scour holes and the deterioration of each. The focus of this overview is the latter. Specifically, the following models have been proposed by Van Noortwijk et al. (Chapter 2).

1.2 The Proposed Models

For the joint distribution of $T_1, T_2, \cdots$, exchangeability and also a " lack of memory property" have been assumed. That is, for any $n = 1, 2, \cdots$, $T_1, T_2, \cdots$, T_n are assumed to be uniform on the simplex $[T_i \geq 0, \sum_{i=1}^{N} T_i]$, where $\sum_{i=1}^{N} T_i$ is assumed known. Then, it follows (Diaconis and Freedman [3]) that the joint density function of $(T_1, \cdots, T_n)$, say $p(T_1, \cdots, T_n)$, is a scale mixture of independent, identically distributed exponential random variables. Furthermore, under the added condition of "posterior linearit" - that is, $E(T_i|T_{i-1}) = c_1 T_{i-1} + c_2$, for some constants $c_1 > 0, c_2 \geq 0$ - the mixing distribution must be a gamma.

If $X(t)$ denotes the cumulative erosion of any scour hole at time $t \geq 0$, then a model for the stochastic behavior of $X(t)$ has been motivated via the following considerations:

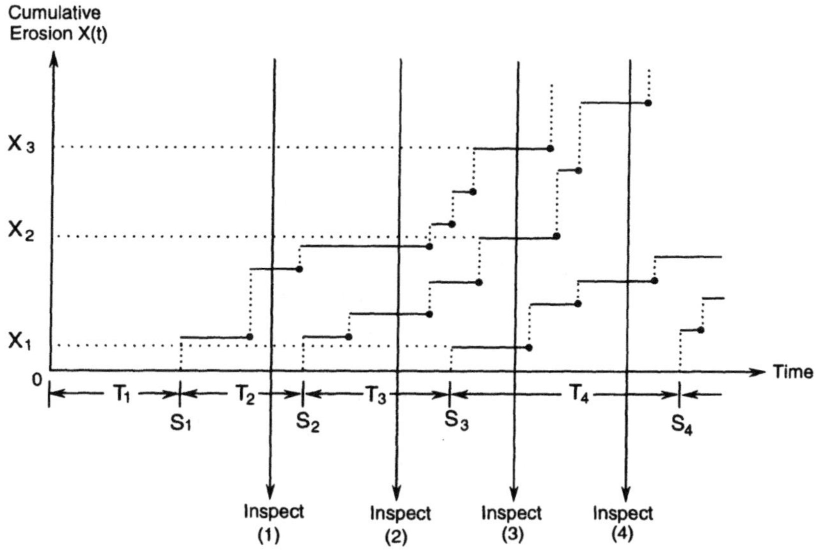

Figure 1 The times of occurrences of scour holes, their erosion, and optimal inspection schedules.

Let $D_i(\tau) \stackrel{\text{def}}{=} X(i\tau) - X((i-1)\tau)$, $\tau > 0$, $i = 1, 2, \cdots$, denote the incremental erosion of a scour hole in the time interval $[(i-1)\tau, i\tau]$; suppose that $X(0) \equiv 0$. Thus, $D_i(\tau) \geq 0$ for all i. Assume that the sequence $\{D_i(\tau); i = 1, 2, \cdots\}$ is an infinite exchangeable sequence of random variables. Furthermore, suppose that for any $n = 1, 2, \cdots$, the vector $[D_1(\tau)/S, D_2(\tau)/S, \cdots, D_n(\tau)/S]$ has a Dirichlet distribution on the simplex $[(D_i(\tau)/S) \geq 0, 1]$ with a parameter vector $[\alpha\tau, \cdots, \alpha\tau]$, where $S = \sum_{i=1}^{n} D_i(\tau)$. Then, for any n, the joint density of $(D_1(\tau), \cdots, D_n(\tau))$, say $\pi(D_1(\tau), \cdots, D_n(\tau))$ is a scale mixture of independent and identically distributed gamma variables with a scale parameter (S/n) and a shape parameter $\alpha\tau$ (Diaconis and Freedman [3]). Furthermore, if posterior linearity of the $D_i(\tau)$'s is desired, then the mixing distribution on (S/n) must be a gamma. The process $\{D_i(\tau); i = 1, 2, \cdots\}$ when constructed as such has been referred to by Van Noortwijk et al. (Chapter 2) as a *generalized gamma process*. This terminology is *not* standard and a purpose of this overview is to provide a streamlined perspective on gamma processes and their generalizations or their extensions. But before we do this, two comments are in order.

The first comment pertains to a motivation for the Dirichlet distribution of the vector $[D_1(\tau)/S, \cdots, D_n(\tau)/S]$. This distribution results if we assume that

$(D_1(\tau)/S)$ has a beta distribution on $[0,1]$ with parameters $(\alpha\tau, \alpha\tau)$, and that the conditional distribution of $(D_2(\tau)/S)$ given $(D_1(\tau)/S)$ is a beta on $[D_1(\tau)/S, 1]$ with parameters $(\alpha\tau, \alpha\tau)$, and so on; that is, the conditional distribution of $(D_3(\tau)/S)$ given $(D_2(\tau)/S)$ is a beta on $[D_2(\tau)/S, 1]$ with parameters $(\alpha\tau, \alpha\tau)$. The second comment pertains to the fact that the scale mixing distributions mentioned above only hold for sequences that are conceptually infinite. Whereas the sequence of T_i's can be conceptually infinite, the sequence of $D_i(\tau)$'s cannot. The sea-bed protection has a finite thickness, say W, and there exists some k, $k < \infty$, such that $\sum_{i=1}^{k} D_i(\tau) = W$. All the same, a theorem of Diaconis and Freedman [3] shows that for a finite exchangeable sequence $\{D_1(\tau), \cdots, D_k(\tau)\}$ the "total variation error" in assuming the proposed (scale mixture) distribution is, for $1 \leq n \leq k-2$, of the order $2(n+1)/(k-n-1)$, which for large k goes to zero. Thus the proposed motivation of the so called "generalized gamma process" for the scour hole erosion phenomenon is not seriously hampered by the finiteness of k.

2 THE GAMMA PROCESS

In its simplest form, a "gamma process" is defined via the following ingredients. First, let $\text{Ga}(\alpha, \beta)$ denote a gamma density with shape α and scale β. That is $\text{Ga}(x|\alpha, \beta) = e^{-\beta x}(\beta x)^{\alpha-1}\beta/\Gamma(\alpha)$. Also, let $\alpha(t); t \geq 0$ be a nondecreasing left continuous real-valued function, with $\alpha(0) \equiv 0$; $\alpha(t)$ is known as the ***shape function***. Define a continuous parameter continuous state stochastic process $\{Z(t); t \geq 0\}$ with the following properties:

1. $Z(0) \equiv 0$;
2. $Z(t)$ has independent increments;
3. for any $t > s$, $(Z(t) - Z(s)) \sim \text{Ga}(\alpha(t) - \alpha(s), 1)$.

Without loss of generality we assume that the sample paths of $\{Z(t); t \geq 0\}$ are left continuous and nondecreasing. Ferguson [5] shows that such a process exists and the process is called a ***gamma process***. Clearly, $Z(t) \sim \text{Ga}(\alpha(t), 1)$, and it can be argued that $\{Z(t); t \geq 0\}$ is a "pure jump process." Thus, its sample path is a step-function (Figure 2).

If the shape function $\alpha(t) = at$, $a > 0$, a constant, then the process $\{Z(t); t \geq 0\}$ has stationary, independent increments; that is, it is a Lévy process. Thus for

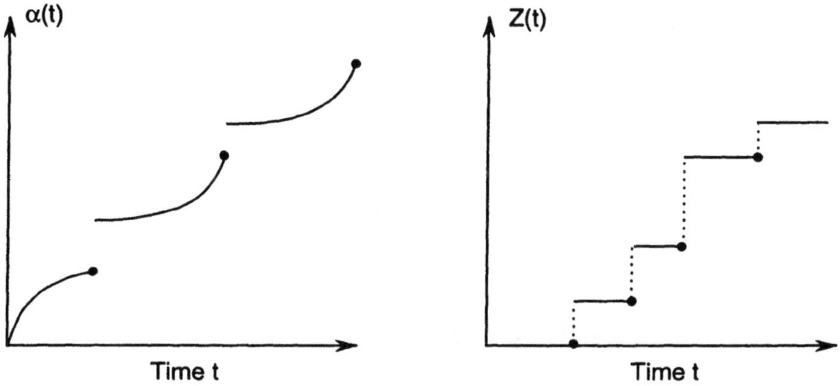

Figure 2 The shape function $\alpha(t)$ and the sample-path $Z(t)$, of a gamma process.

any $u > 0$, and any $t \geq 0$, $(Z(t+u) - Z(t))$ has a gamma distribution with shape αu. Also, if $A = \{t; \alpha(t) \neq \alpha(t^-)\}$ is the set of discontinuities of $\alpha(t)$, then for $t \in A$, the process $\{Z(t); t \geq 0\}$ has a jump at t (a. s.) and the size of the jump has a gamma distribution with shape $\alpha(t) - \alpha(t^-)$ and scale 1. Finally, it can be shown that for $0 \leq s \leq t$, and all pairs s and t:

$$E[Z(t) - Z(s)] = \int_{(s,t]} d\alpha(u),$$

$$V[Z(t) - Z(s)] = \int_{(s,t]} d\alpha(u); \quad \text{see Dykstra and Laud [4],}$$

where E denotes expectation and V denotes variance.

3 THE GENERALIZED GAMMA PROCESS

The generalized gamma process results from the following shape transformation of the gamma process. Let $\{Z(t); t \geq 0\}$ be a gamma process with a shape function $\alpha(t)$. Let $a(t)$ and $b(t)$ be any two increasing functions. Define

$$T_{(a,b)}(Z(t)) = \int_{(0,a(t)]} \frac{1}{b(u)} dZ(u).$$

Then $T_{(a,b)}(Z(t))$ is a process that is obtained by changing the scale of $Z(t)$ by $b(t)$ and changing time via $a(t)$. The process $\{T_{(a,b)}(Z(t)); t \geq 0\}$ is called a *generalized gamma process* because of the following theorem [2].

Theorem 5 (Çinlar, 1980) *If $a(t)$ is continuous and $a(0^+) = 0$, then $\{T_{(a,b)}(Z(t)); t \geq 0\}$ is a gamma process with a shape function $(\alpha.a)$ and scale $(b.a)$.*

3.1 The Extended Gamma Process

Let $\beta(t), t \geq 0$ be a positive right continuous real valued function, bounded away from 0, with left-hand limits that exist. Let $\{Z(t); t \geq 0\}$ be a gamma process with a shape function $\alpha(t)$ and a scale 1. Let

$$r(t) = \int_{[0,t)} \beta(u)\, dZ(u).$$

Then Dykstra and Laud [4] refer to the process $\{r(t); t \geq 0\}$ as an extended gamma process. The process has independent increments, and is a pure jump process. The distribution function of $r(t)$ is intractable, but the following properties hold:

1. $E(r(t)) = \int_{[0,t)} \beta(s)\, d\alpha(s)$;
2. $V(r(t)) = \int_{[0,t)} \beta^2(s)\, d\alpha(s)$;
3. The characteristic function of $r(t)$ in some neighborhood of θ is

$$\psi_{r(t)}(\theta) = \exp\left[-\int_{[0,t)} \ln(1 - i\beta(s)\,\theta)\, d\alpha(s)\right].$$

4 THE LÉVY DECOMPOSITION OF GAMMA PROCESSES

If $\alpha(t)$ is continuous then the gamma process has no points of discontinuity and it can be represented as the sum of a countable number of jumps of random height at a countable number of random points; see Ferguson and Klass [6]. Furthermore, for any fixed $\epsilon > 0$, if $N(t,\epsilon)$ denotes the number of jumps in $(0,t)$ of magnitude greater than or equal to ϵ, then $\{N(t,\epsilon); t \geq 0\}$ is a Poisson process with an intensity function $M(t,\epsilon)$, where

$$M(t,\epsilon) = p(t) \int_{\epsilon}^{\infty} \frac{1}{u} e^{-u}\, du,$$

with $p(t)$ depending on $\alpha(t)$. When $\alpha(t) = at$, then $p(t) = p$ and $M(t,\epsilon)$ is independent of t; see Basawa and Prakasa Rao [1], p. 105.

Also, if $u(\epsilon)$ is the size of a jump, given that it is bigger than $\epsilon > 0$, then when $\alpha(t) = at$, $U(\epsilon)$ has a truncated gamma distribution with shape 0 and truncation point ϵ.

5 REPRESENTATION OF A GAMMA PROCESS

A result due to Wiener is that the Brownian motion on $[0,\pi]$ has a representation as a countable sum of the form

$$W(t) = tY_0 + \sqrt{2} \sum_{m=1}^{\infty} Y_m \frac{\sin(mt)}{m},$$

where $Y_0, Y_1, \cdots,$ are independent and identically distributed as a Gaussian distribution with mean 0 and variance 1. We seek an analogous representation for a process with independent increments having no Gaussian components and no fixed points of discontinuity – such as a gamma process. It is known that such processes have infinitely divisible distributions for which the log characteristic function is of the form

$$\psi_t(u) = \int_0^{\infty} \left(e^{iuz} - 1 - \frac{iuz}{1+z^2} \right) dN_t(z),$$

where the Lévy measure N_t on the Borel sets of $(0,\infty)$ is nondecreasing and continuous in $t \in [0,1]$ and has the following properties:

1. $N_0 \equiv 0$;
2. For every Borel set B, $N_{t_1}(B) \leq N_{t_2}(B)$, for $0 \leq t_1 \leq t_2 \leq 1$;

3. $\int_o^\infty \frac{z}{1-z^2}\, dN_1(z) < \infty$, where $N_1(z) = -N_1[z, \infty)$.

The jumps of the independent increments process defined above are all positive. Let $J_1 \geq J_2 \geq \cdots$, be the ordered heights of the jumps, and let T_j be the time at which J_j occurs. Then,

$$P(J_1 \leq x) = e^{N_1(x)}, \text{and}$$

$$P(J_1 \leq x_j | J_{j-1} = x_{j-1}, \cdots, J_1 = x_1) = e^{N_1(x_j) - N_1(x_{j-1})}, \quad 0 < x_j < x_{j-1}.$$

Condition 2 above implies that N_{t_1} is absolutely continuous with respect to N_{t_2}, whenever $t_1 \leq t_2$. Thus, the Radom-Nikodym derivative of N_t with respect to N_1, say $n_t(z)$ exists, where

$$n_t(z) = \frac{dN_t}{dN_t}(z), \quad \forall z \in (0, \infty).$$

Also, $n_t(z)$ is a nondecreasing function on $[0, 1]$ with $n_0(z) \equiv 0$, and $n_1(z) = 1$. The distribution of the points $T_1, T_2, \cdots$, given $J_1, J_2, \cdots$, is essentially the same as the distribution of independent random variables having distribution $n_t(J_1), n_t(J_2), \cdots$; see Ferguson and Klass [6]. We now have the representation theorem,

Theorem 6 (Ferguson and Klass (1972))

$$X(t) = \sum_{j=1}^{\infty} J_j I_{[0, n_t(J_j))}(U_j) - C_j(t),$$

where $U_1, U_2, \cdots$, *are independent and identically uniformly distributed on* $[0, 1]$, *independent of* $J_1, J_2, \cdots$, *and* $C_j(t)$ *is a centering constant*

$$C_j(t) = \int_{j-1}^{j} \frac{N_1^{-1}(-v)}{1 + (N_1^{-1}(-v))^2} n_t(N_1^{-1}(-v))\, dv.$$

If the Lévy measure $N_t(z) = G(t)N_1(z)$ for some nondecreasing function $G(t)$ on $[0, 1]$ is such that $G(0) \equiv 0$ and $G(1) \equiv 1$, then $X(t)$ has the representation

$$X(t) = \sum_{j=1}^{\infty} J_j I_{[0, G(t))}(U_j) - G(t) C_j, \text{where}$$

$$C_j = \int_{j-1}^{j} \frac{N_1^{-1}(v)}{1 + (N_1^{-1}(-v))^2}\, dv.$$

The above results when specialized for the case of a gamma process yield the following representation.

Recall that the characteristic function of a gamma distribution with shape α and scale 1 is of the form

$$\exp\left[\int_0^\infty \left(e^{iux-1}\right)\, dN_1(x)\right],$$

where $N_1(x) = -\alpha \int_x^\infty e^{-y}y^{-1}\, dy$. Then the representation

$$X(t) = \sum_{j=1}^{\infty} J_j I_{[0,G(t))}(U_j)$$

converges, with probability 1, to a gamma process with a shape function $\alpha G(t)$ and scale 1.

An alternate representation of the gamma process is in a preprint by Van der Weide (1995) of the Technical University of Delft, Netherlands. Specifically, he represents a gamma process as integrals of Poisson processes.

REFERENCES

[1] I. V. Basawa and B. L. S Prakasa Rao. *Statistical Inference for Stochastic Processes.* Academic Press, London, 1980.

[2] E. Çinlar. On a generalization of gamma processes. *J. Appl. Prob.*, 17:467–480, 1980.

[3] P. Diaconis and D. A. Freedman. A dozen de finetti-style results in search of a theory. *Ann. Inst. Henri Poincare*, 23:397–423, 1987.

[4] R. L. Dykstra and P. Laud. A bayesian nonparametric approach to reliability. *Ann. Statist*, 9:356–367, 1981.

[5] T. S. Ferguson. Prior distributions on spaces of probability measures. *Ann. Statist.*, 2:614–629, 1973.

[6] T. S. Ferguson and M. J. Klass. A representation of independent increment processes without gaussian components. *Ann. Math. Statist.*, 43:1634–1643, 1972.

6

GAMMA PROCESSES

Hans van der Weide

Department of Mathematics,
Delft Unviversity of Technology, Delft, The Netherlands

ABSTRACT

In this paper we will construct gamma processes from Poisson processes and give a description of some of their properties.

1 INTRODUCTION

A random variable X has a gamma distribution with shape parameter $a > 0$ and scale parameter $b > 0$ ($\Gamma(a,b)$ distribution) if its probability density function is given by:

$$f_{a,b}(x) = \frac{b^a}{\Gamma(a)} x^{a-1} e^{-bx} I_{(0,\infty)}(x).$$

We recall that the moment generating function of the $\Gamma(a,b)$ distribution is

$$\int_0^\infty e^{-\lambda x} f_{a,b}(x)dx = (\frac{b}{b+\lambda})^a \text{ for } \lambda > -b.$$

So if X is $\Gamma(a,b)$ distributed then bX is $\Gamma(a,1)$ distributed. Note that $\mathbf{E}(bX) = \sigma^2_{bX} = a$. Finally note that if X_i is $\Gamma(a_i, b_i)$ distributed, $i = 1,2$, and if X_1 and X_2 are independent, then the sum $X_1 + X_2$ is gamma distributed if and only if $b_1 = b_2$.

A (real-valued) stochastic process $Z = (Z(t); t \geq 0)$ is called a gamma process if $Z(0) = 0$ and if it has independent, gamma distributed increments, i.e. $Z(t) - Z(s)$ is $\Gamma(t-s, 1)$-distributed for $s < t$. The gamma process is a special case of what is called a homogeneous differential process in [2]. We will construct

R. Cooke et al. (eds.), Engineering Probabilistic Design and Maintenance for Flood Protection, 77–83.

a version of the gamma process with increasing, right continuous realisations. Then for $\lambda > 0$

$$\mathbf{E}[e^{-\lambda Z(t)}] = e^{-t\psi(\lambda)},$$

where

$$\psi(\lambda) = m\lambda + \int_{(0,\infty)} [1 - e^{-\lambda u}]\nu(du)$$

$$m \geq 0, \int_{(0,\infty)} [1 - e^{-u}]\nu(du) < +\infty;$$

$\nu(du)$ is the Lévy measure of the process. Corresponding to this formula, Lévy gave a decomposition of the sample path into a linear part plus an integral of differential Poisson processes:

$$Z(t) = mt + \int_{(0,\infty)} l\mathcal{P}([0,t] \times du),$$

$\mathcal{P}(dt \times du)$ being the Poisson measure with mean $dt \otimes \nu(du)$. For a gamma process we have

$$\mathbf{E}(e^{-\lambda Z(t)}) = (\frac{1}{\lambda + 1})^t = e^{-t\ln(\lambda+1)}.$$

Hence the Lévy measure satisfies the equation:

$$\int_{(0,\infty)} [1 - e^{-\lambda u}]\nu(du) = \ln(\lambda + 1).$$

But

$$\begin{aligned} \int_{(0,\infty)} [1 - e^{-\lambda u}]\nu(du) &= \int_{(0,\infty)} \nu(du) \int_0^u \lambda e^{-\lambda x} dx \\ &= \int_0^\infty \lambda e^{-\lambda x} \nu(x, \infty) dx. \end{aligned}$$

Hence

$$\int_0^\infty e^{-\lambda x} \nu(x, \infty) dx = \frac{\ln(\lambda + 1)}{\lambda}$$

from which we conclude that $\nu(x, \infty) = \int_x^\infty y^{-1} e^{-y} dy$, so the Lévy measure for the gamma process is given by

$$\nu(du) = \frac{e^{-u}}{u} du.$$

In this paper we will try to describe gamma processes as integrals of Poisson processes.

Let $\mathbf{P}_n$ be the distribution of a Poisson process on the polish space E with intensity n. This means that $\mathbf{P}_n$ is a probability measure on the space $\Omega = \mathcal{M}^\bullet(E)$ of simple point measures, i.e. measures ω on E which can be represented as locally finite sums of unit-pointmasses δ_x, $x \in E$:

$$\delta_x(A) = \begin{cases} 1 & \text{if } x \in A \\ 0 & \text{otherwise.} \end{cases}$$

A point measure ω is called simple if $\omega(\{x\}) = 0$ or 1 for every $x \in E$. For measurable sets $A \subset E$, let $N(A)$ denote the random variable $\omega \in \Omega \mapsto \omega(A)$. With respect to $\mathbf{P}_n$, the random variables $N(A)$ are Poisson distributed with mean $n(A)$:

$$\mathbf{P}_n(N(A) = k) = \frac{n(A)^k}{k!} e^{-n(A)}, \ k = 0, 1, 2, \ldots$$

Further, the random variables $N(A)$ and $N(B)$ are independent for disjoint sets $A, B \subset E$. The characteristic functional of $\mathbf{P}_n$ is given by

$$\int_{\mathcal{M}^\bullet(E)} e^{-\int f(x)\omega(dx)} \mathbf{P}_n(d\omega) = e^{-\int [1-e^{-f}]dn}, \tag{6.1}$$

f being a non-negative measurable function on E. Taking $f = 1_A$, this is a consequence of the Poisson-distribution of $N(A)$. The distribution $\mathbf{P}_n$ is completely determined by its characteristic functional. For further properties of Poisson point processes we refer to [1], [3] and [4].
Let $\phi : E \mapsto E$ be a measurable transformation such that ϕn is a diffuse measure on E, where ϕn is the image measure defined by

$$\phi n(A) = n(\phi^{-1}A).$$

Let $\Phi : \mathcal{M}^\bullet(E) \mapsto \mathcal{M}^\bullet(E)$ be the map defined by $\Phi(\omega) = \phi\omega$. It follows that for a non-negative measurable function f on E we have

$$\int_E f(x)\Phi\omega(dx) = \int_E f \circ \phi(x)\omega(dx).$$

Then the image measure $\Phi\mathbf{P}_n$ is the distribution of a Poisson process on E with intensity ϕn.

2 CONSTRUCTION OF GAMMA PROCESSES

Let $T = [0, \infty)$, $U = (0, \infty)$ and $E = T \times U$, all with the usual Borel σ-algebras. Denote the Lebesgue measure on T by dt. Let ν be a Borel measure on U such that

$$\int_U [1 - e^{-u}]\nu(du) < \infty. \tag{6.2}$$

Let $\Omega = \mathcal{M}^\bullet(E)$ with the usual σ-algebra $\mathcal{B}$ and let $\mathbf{P}_{dt\otimes\nu}$ be the Poisson process on E with intensity $dt \otimes \nu$. The triple $(\Omega, \mathcal{B}, \mathbf{P}_{dt\otimes\nu})$ is a probability space. Define for $\omega \in \Omega$ and $t \in T$

$$X_t(\omega) = \int_E \omega(d\sigma du) I_{[0,t]}(\sigma)u. \tag{6.3}$$

Then X_t is a random variable on $(\Omega, \mathcal{B}, \mathbf{P}_{dt\otimes\nu})$. Let $\mathbf{E}$ denote expectation with respect to $\mathbf{P}_{dt\otimes\nu}$. Then the moment generating function of X_t is given by

$$\begin{aligned} \mathbf{E}(e^{-\lambda X_t}) &= \int \mathbf{P}_{dt\otimes\nu}(d\omega) e^{-\int_E \omega(d\sigma du) I_{[0,t]}(\sigma)\lambda u} \\ &= \exp[-\int \nu(du) \int_0^\infty d\sigma(1 - e^{-I_{[0,t]}(\sigma)\lambda u})] \\ &= \exp[-t \int \nu(du)(1 - e^{-\lambda u})]. \end{aligned}$$

Here the second line is an application of (1). Using condition (2), it follows that

$$\mathbf{P}_{dt\otimes\nu}(X_t < \infty) = \lim_{\lambda \downarrow 0} \mathbf{E}(e^{-\lambda X_t}) = 1,$$

so the process X is defined $\mathbf{P}_{dt\otimes\nu}$-a.s.. It is clear that the process $X = (X_t; t \geq 0)$ has independent increments and that $X_0 = 0$ $\mathbf{P}_{dt\otimes\nu}$-a.s.. The sample path's of X are increasing and right continuous (by Lebesgue's dominated convergence theorem). Taking $\nu(du) = u^{-1}e^{-u}du$, the process X is the gamma process described in the introduction. Note that we have in this case $\mathbf{E}(X_t) = t$. The function $a(t) = \mathbf{E}(X_t)$ is called the shape function of the gamma process.
We will continue with a more general process by considering other shape functions a. Let $a : T \mapsto T$ be an non-decreasing, right continuous function with $a(0) = 0$. A process $Z = (Z_t; t \geq 0)$ is called a gamma process with shape function a, if $Z_0 = 0$ and if Z has independent increments $Z_t - Z_s, t > s$, which are $\Gamma(a(t) - a(s), 1)$ distributed. To construct such a process, define

$$Z_t(\omega) = \int_E \omega(d\sigma du) I_{[0,t]}(\sigma)u.$$

on the probability space $(\Omega, \mathcal{B}, \mathbf{P}_{da(t)\otimes\nu})$ with $\nu(du) = u^{-1}e^{-u}du$. We get

$$\mathbf{E}(e^{-\lambda Z_t}) = \exp[-\int \nu(du) \int_0^\infty da(\sigma)(1 - e^{-I_{[0,t]}(\sigma)\lambda u})]$$

$$= \exp[-a(t)\int \nu(du)(1-e^{-\lambda u})]$$
$$= (\frac{1}{\lambda+1})^{a(t)},$$

from which we can conclude that Z is a gamma process with shape function a. It is clear that distributions of the processes Z and $\{X_{a(t)}; t \geq 0\}$, where X is the gammaprocess with shape function $a(t) = t$, are the same. Note that if $\Delta a(t) = a(t) - a(t-) > 0$, then ΔZ_t is $\Gamma(\Delta a(t), 1)$ distributed. So Z has almost surely jumps at discontinuity points of the shape function a.

3 THE EXTENDED GAMMA PROCESS

Let $Z = (Z_t; t \geq 0)$ be the gamma process with shape function a as constructed in section 2. Let $\beta : T \mapsto T$ be some positive, right continuous measurable map. Define a new random process $Y = (Y_t; t \geq 0)$ by:

$$Y_t = \int_0^t \beta(s) Z_{ds}.$$

Using the point process representation for Z:

$$Y_t(\omega) = \int_E \omega(d\sigma, du)\beta(\sigma) u I_{[0,t]}(\sigma).$$

¿From this formula it is clear that the process Y is a pure jump process with independent increments. The process Y is called an extended gamma process if the factor ν in the intensity measure of the Poisson process $\mathcal{P}_{da(t)\otimes\nu}$ is chosen to be the Lévy measure of the gamma process: $\nu(du) = u^{-1}e^{-u}du$. The moment generating function of Y_t is given by:

$$\begin{aligned}
\int_\Omega e^{-\lambda Y_t(\omega)} \mathbf{P}_{da(t)\otimes\nu}(d\omega) &= \int \mathbf{P}_{da(t)\otimes\nu}(d\omega) e^{-\int_E \omega(d\sigma du) I_{[0,t]}(\sigma)\lambda u \beta(\sigma)} \\
&= \exp[-\int \nu(du) \int_0^\infty da(\sigma)(1 - e^{-I_{[0,t]}(\sigma)\lambda u \beta(\sigma)})] \\
&= \exp[-\int_0^t da(\sigma) \int (1 - e^{-\lambda u \beta(\sigma)}) \frac{e^{-u}}{u} du] \\
&= \exp[-\int_0^t \ln(1 + \lambda\beta(\sigma)) da(\sigma)].
\end{aligned}$$

By differentiation we can get expressions for the mean and the variance of Y_t:

$$\mathbf{E}(Y_t) = \int_0^t \beta(\sigma)da(\sigma) \text{ and } Var(Y_t) = \int_0^t \beta^2(\sigma)da(\sigma).$$

There is an alternative way to construct extended gamma processes. The formula

$$Y_t(\omega) = \int_E \omega(d\sigma, du)\beta(\sigma)uI_{[0,t]}(\sigma)$$

suggests that we should consider the following transformation of the space E:

$$\phi : (\sigma, u) \mapsto (\sigma, \beta(\sigma)u).$$

The image of the intensity measure $u^{-1}e^{-u}da(t) \otimes du$ under this map is the diffuse measure

$$n = u^{-1}e^{-\frac{u}{\beta(\sigma)}}da(t) \otimes du.$$

It follows that we can construct the extended gamma process Y as

$$Y_t(\omega) = \int_E \omega(d\sigma, du)uI_{[0,t]}(\sigma)$$

on the probability space $(\Omega, \mathcal{B}, \mathbf{P}_n)$.

4 SOME PROPERTIES

Let X be a gamma process with shape function a. Let $\epsilon > 0$ be given. The number of jumps $> \epsilon$ during the time interval $[s,t]$, is Poisson distributed with mean $(a(t) - a(s-)) \int_\epsilon^\infty \frac{e^{-u}}{u}du$.
Define for $s < t$

$$m_{[s,t]} = \sup\{\Delta X_r \mid s \leq r \leq t\}.$$

Then, using the representation of section 2, we have $m_{[s,t]}(\omega) \leq h$ if and only if $\omega([s,t] \times (h,\infty)) = 0$. It follows from the properties of the Poisson process that

$$\mathbf{P}_{da(t)\otimes\nu}(m_{[s,t]} \leq h) = e^{-(a(t)-a(s-))\int_h^\infty \frac{e^{-u}}{u}du}.$$

We continue with properties of the crossingtimes. Define for $l > 0$

$$\tau_l = \inf\{t \mid X_t \geq l\}.$$

It is clear that

$$\mathbf{P}_{da(t)\otimes\nu}(\tau_l \leq t) = \mathbf{P}_{da(t)\otimes\nu}(X_t \geq l) = \int_l^\infty \frac{1}{\Gamma(a(t))}x^{a(t)-1}e^{-x}dx.$$

Using the representation of section 2 we calculate now the double Laplace-transform.

$$\begin{aligned}
&\int_0^\infty \mathbf{E}(e^{-\lambda\tau_l})e^{-\mu l}dl \\
&= \int \mathbf{P}_{da(t)\otimes\nu} \int \omega(d\sigma du) \int_0^\infty I_{(0,u]}(l - X(\sigma-,\omega))e^{-\lambda\sigma}e^{-\mu l}dl \\
&= \int \mathbf{P}_{da(t)\otimes\nu} \int \omega(d\sigma du)e^{-\lambda\sigma}\frac{1}{\mu}e^{-\mu X(\sigma-,\omega)}(1 - e^{-\mu u}) \\
&= \frac{1}{\mu}\int_0^\infty da(\sigma)\int \nu(du)e^{-\lambda\sigma}(1 - e^{-\mu u})(\mu + 1)^{-\sigma} \\
&= \frac{\ln(\mu + 1)}{\mu}\int_0^\infty e^{-(\lambda+\ln(\mu+1))\sigma}da(\sigma).
\end{aligned}$$

Using the second construction of section 3, we can derive formulas for an extended gamma process Y in an analogous way. For example, the number of jumps $> \epsilon > 0$ during the time interval $[s,t]$ is Poisson distributed with mean $n([s,t] \times (\epsilon,\infty)) = \int_s^t \beta(\sigma)da(\sigma)\int_\epsilon^\infty \frac{e^{-u}}{u}du$.

REFERENCES

[1] Daley, D.J. and Vere-Jones, D. (1988). *An Introduction to the Theory of Point Processes.* Springer, New York.

[2] Ito, K. and McKean, H.P. (1965). *Diffusion Processes and their Sample Paths.* Springer-Verlag, Berlin.

[3] Kingman, J.F.C.(1993). *Poisson Processes.* Clarendon Press, Oxford.

[4] van der Weide, J.A.M. (1994). *Stochastic Processes and Point Processes of Excursions.* CWI, Amsterdam.

7

REPLY TO THE REVIEWS

Jan van Noortwijk*, Matthijs Kok, and Roger Cooke***

**Department of Mathematics, Delft University of Technology, Delft, The Netherlands*
***Delft Hydraulics, De Voorst, The Netherlands*

First, I would like to thank the three discussants for their interesting comments, questions, and suggestions.

Leo Klatter points to a very important aspect in the modelling of maintenance: the communication between the maintenance engineers and the mathematicians. To assure successful applications, a maintenance model has to fullfill almost conflicting requirements: on the one hand, it should describe the reality well; on the other hand, it should not demand data which is not available. It is not surprising that the few maintenance models that have been successfully applied are the most simple ones: viz. the well-known age and block replacement models (see Dekker [8]). Therefore, the approximations that have been made for modelling the maintenance of the barrier's sea-bed protection should be viewed in the light of the availability of deterioration data.

Nozer Singpurwalla has written an excellent overview of gamma processes and their generalisations. The overview summarises the nice mathematical properties of gamma processes. Indeed, we must admit that the generalised gamma processes which were studied in the main paper can better be denoted by *scale mixtures of gamma processes.*

John Shortle raises several interesting issues concerning the physics of the initiation and development of scour holes. The main reason for considering the initiation of scour holes as a mixture of stationary (homogeneous) Poisson processes is the current lack of data (so far, 1996, no scour holes have been detected). In the event that 12 scour holes are observed in 1996-1997, more data will have become available and the process of occurrence of scour holes may be regarded as a mixture of non-stationary (non-homogeneous) Poisson processes. In this

R. Cooke et al. (eds.), Engineering Probabilistic Design and Maintenance for Flood Protection, 85–86.

case, Silver's model can best be applied to adjust the maintenance model for the block mats (specifically, by replacing λ, the average inter-occurrence time of scour holes, with λ_i, the average inter-occurrence time of scour holes in unit time i, $i \in \mathbb{N}$). In fact, a non-homogeneous Poisson process can be transformed into a homogeneous Poisson proces by transforming time, which should bring us back to the original situation. It should be stressed, however, that there is no evidence that the block mats have a lifetime of only 10 years!

John Shortle's second suggestion addresses the development of scour holes: whether the rate of scour erosion is constant or decreases over time according to an empirical power law between the depth of a scour hole and the time after its initiation. In fact, we have used this power law to estimate the average rate of erosion for scour holes with a depth lying in between the detectability level (2 m) and the failure level (15 m). Note that the uncertainty in the value of the power in the empirical relation is rather large: extrapolating results from a small laboratory set-up to the large Eastern-Scheldt barrier induces many uncertainties. According to the empirical power law, scour holes cannot fill back up, have no maximum depth, and increase over time (also for the block mats).

PART II

8

A PHYSICS-BASED APPROACH TO PREDICTING THE FREQUENCY OF EXTREME RIVER LEVELS

Steve Chick, John Shortle*, Pieter van Gelder and Max Mendel***

Department of Industrial and Operations Engineering, University of Michigan, Ann Arbor, Michigan 48109
**Department of Industrial Engineering and Operations Research, University of California, Berkeley, California 94720*
***Department of Civil Engineering, Delft University of Technology, Delft, The Netherlands*

ABSTRACT

A new model for predicting the frequency of extreme river levels is proposed which encapsulates physical knowledge about river dynamics, including the Chezy equation for discharge and water input from tributaries. A simple example shows how changes in discharge characteristics changes the extreme river level frequencies. Solutions are provided for special cases, and directions for more general techniques are provided.

1 INTRODUCTION

Two major purposes for flood prediction are (1) short-term evacuation purposes and (2) long-term flood frequency analysis for flood defense system design. At present, the analysis for each purpose is quite different. Short-term prediction uses extensive physical knowledge of continuous-time river catchment properties, including river dynamics, prior and predicted weather over the catchment, and run-off models. The FLOFOM model described by Berger [1] and used by the Rijkswaterstaat (Dutch Ministry of Transport and Water Defense) to analyze the Maas is an example of such a model. Long-term flood prediction, on the other hand, has traditionally fit peak flood levels to statistical models such as the log-Pearson III [2] or Gumbel [3]. For example, Fig. 1 is a statistical fit of flood data from the Rhine [4]. Cunnane [5] provides a survey of

R. Cooke et al. (eds.), Engineering Probabilistic Design and Maintenance for Flood Protection, 89–107.

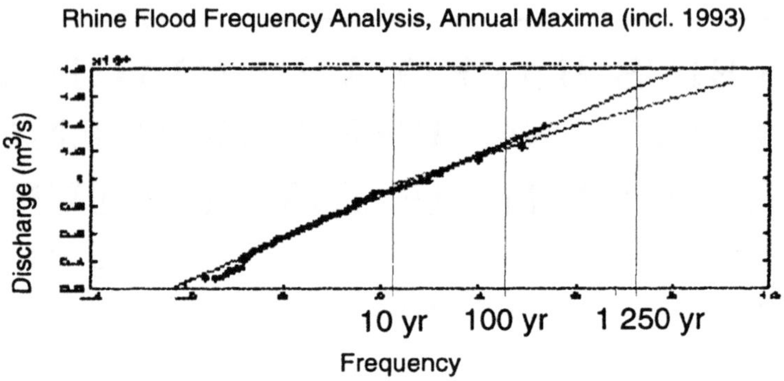

Figure 1 Empirical river discharge frequencies of the Rhine and fitted distributions.

these and other statistical models and indicates that goodness of fit tests (χ^2, Kolmogorov-Smirnov, moment-ratio diagrams) for selecting a particular model are inconclusive. Difficulty in comparing hydrological parameters (run-off coefficients, Chezy's equation discharge parameters) with statistical parameters (slope, intercept) is one contributor to this inconclusiveness.

In this paper, a new model for predicting the frequency of extreme river levels is proposed which encapsulates physical knowledge about river dynamics, including the Chezy equation for discharge and water input from tributaries. The resulting model therefore makes an explicit connection between hydrological and statistical parameters. The model accounts for the river dynamics at a given location by modeling both how water gets into the river (via upstream tributaries) and how water leaves (discharge modeled by Chezy's equation). A simplified physical model is presented in Sec. 2. The mathematics needed for calculating long-term flood frequencies is presented in Sec. 3. Calculations for a simple test case are given in Sec. 4, where it is shown how changes in the Chezy equation parameter directly change the water level for floods with the same return period. Solution techniques for the general case are presented in Sec. 5.

We show that the parameters of the Chezy equation directly affect the shape of the curve which relates flood frequency and flood volume. Variation in the shape of this curve is expected for various values of the Chezy parameters. The usual statistical models have a particular shapes which are not related to the Chezy parameters. (For instance, the Gumbel model assumes a linear relation on log paper).

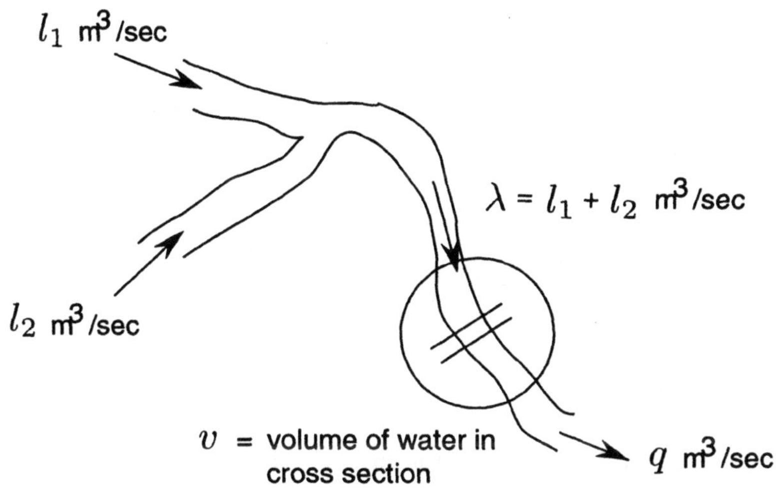

Figure 2 A simplified physical model for the river. At the given cross-section, the river level is determined by input from $K = 2$ major tributaries and the rate of discharge.

2 PHYSICAL MODEL FOR RIVER SYSTEM

Short-term flood prediction makes extensive use of physical modeling of the river system. We seek to use a physical model for the river system as part of the long-term prediction model. The main components of a physical model of a river at a particular cross-section are descriptions of the discharge rate, and the input rate of water from upstream sources (Fig. 2). Because these models may be complex, we propose the following simplified model for how water enters and leaves a river in order to better understanding the viability of a physical-model based approach to long-term flood prediction. The physical model presented here is used to calculate the frequency of floods in later sections.

The rate of discharge from a thin cross-section of a river is a function of the geometry of the cross-section, and the volume of water v in that cross-section. Thus, the discharge $q(v)$ has a functional form which is different for the many geometries encountered in practice, such as those seen in Fig. 3.

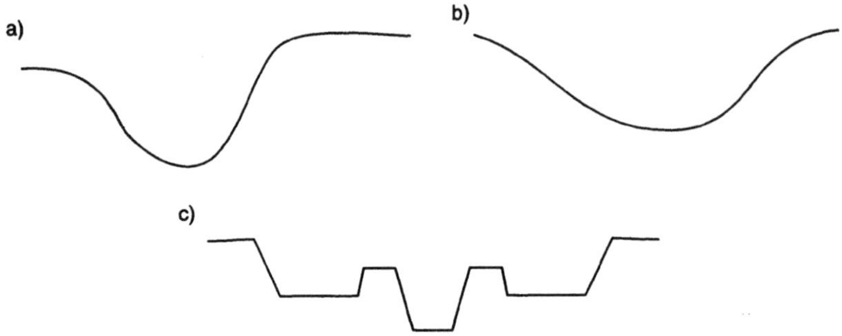

Figure 3 Different geometries for river cross sections.

A standard, accepted description of the rate of discharge is given by Chezy's equation,

$$q(v) = av^b \tag{8.1}$$

for some constants a and b, and q has units $\mathrm{m}^3/\mathrm{sec}$.[1] We will use the Chezy equation to describe discharge in this paper, although all the results are valid for more general functions for discharge as well.

The other major feature of a physical model, the rate of water input, is determined by the cumulative effect of upstream tributaries. Suppose that there are K major tributaries whose catchments affect the cross-section under consideration, and that the flow time from tributary i to the river cross-section in question is approximately δ_i. Then

$$\lambda(t) \approx \sum_{i=1}^{K} l_i(t - \delta_i)$$

where $\lambda(t)$ is the rate of water input, $l_i(t)$ is the discharge in $\mathrm{m}^3/\mathrm{sec}$ from tributary i at time t (Fig. 2).

Short term prediction models generally have sophisticated models for relating the weather, runoff, and other factors to l_i. The number of parameters used by short-term models is quite large. For purposes of illustrating ideas, a simplified model is presented. We model the state of the tributaries as a Markov-modulated fluid source (MMFS). The MMFS is commonly used to gain insights into information flow in high-speed communications networks.[6]

[1] Many developments have q as a function of the river height h. The present formulation is more convenient for the development below.

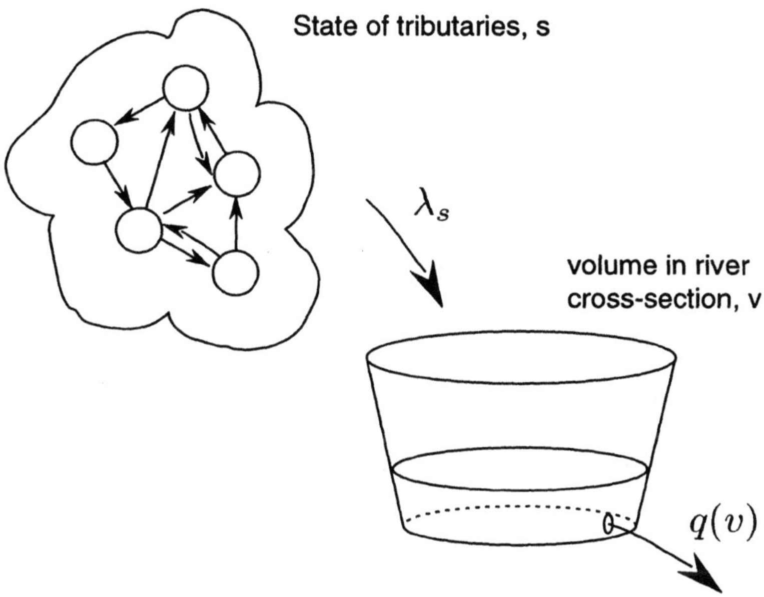

Figure 4 Conceptual model for river: The state of the tributaries s is modeled by a continuous time Markov chain (CTMC). The CTMC state determines the rate of input λ_s from tributaries. Chezy's equation for discharge $q(v)$.

The basic components of the MMFS are the state and fluid rate. In the present work, the state s is defined to be the array of discharges from the K tributaries. The fluid rate is defined to be the sum of the discharges from the tributaries, and is the input rate of water into the river cross-section.

We make a necessary generalization for the MMFS in order to apply it to the flood-frequency problem. In communications, the MMFS generally assumes a constant bit rate, with the choice $q(v)$ = constant bit rate. The present work makes an extension by permitting $q(v)$ to be a non-constant function, such as Chezy's equation.

The use of the MMFS to model the river system requires two simplifications of the physical model. First, the MMFS that the l_i form a discrete set, $l_i \in \mathcal{L}$. For a particular river system, one might set $\mathcal{L} = \{0, 25, 50, \ldots, \bar{q}\}$ (increments not necessarily constant). The maximal discharge $\bar{q}$ is chosen to be greater than the largest imaginable discharge from the tributaries (this insures the number of states is finite). With this assumptions, the effect of river catchment at time t on the river cross-section is described by $S(t) = (l_1(t-\delta_1), \ldots, l_K(t-\delta_K)) \in \mathcal{L}^K$. We call $S(t)$ the state of the tributaries at time t.

Set $N = |\mathcal{L}|^K$ to be the number of possible states of the tributaries. Number the states from $s = 1, \ldots, N$. For each state $s \in \mathcal{L}^K$, the rate of water input λ_s into the river cross-section is

$$\lambda_s = \sum_{i=1}^{K} l_i$$

Set $\Lambda = \text{diag}(\lambda_s)$ to be the diagonal matrix of discretized river inputs.

A second assumption of the MMFS is that the random state transitions are memoryless. Equivalently, the catchment is modeled as a continuous-time Markov chain (see, eg [7]), with state transition matrix

$$M = [\mu_{ij}] \tag{8.2}$$

with μ_{ij} equal to the state transitions rate from state i to j (when $i \neq j$), and μ_{ii} the negative of the rate out of state i.

$$\mu_{ij} = \begin{cases} \lim_{h\to 0} \text{Prob}\left(S(t+h) = j \mid S(t) = i\right)/h & i \neq j \\ -\sum_{k\neq i} \mu_{ik} & i = j \end{cases} \tag{8.3}$$

where $i, j \in \mathcal{L}^K$ and $S(t)$ is the state of the tributaries at time t. Thus, the rows add to 0. Under mild technical conditions, the long run probabilities π_i of being in state i satisfy

$$\vec{\pi} M = 0$$

Although M is a large matrix of size $N \times N$, it is sparse. This is due to the fact that rivers do not skip discharge levels instantaneously. For example, with discrete values $25, 50, 75$ m^3/sec, a river cannot skip 50 when changing from 25 to 75 m^3/sec. There are at most K ways to have an increased flow (an increase in one of the K tributaries input flows) and at most K ways to decrease. Thus, each row will have a maximum of $2K+1$ non-zero entries, including the diagonal entry.

This model provides sufficient mathematical structure to calculate the frequency of extreme floods, as detailed further in Sec. 3

3 FLOOD FREQUENCY CALCULATIONS

The mathematics needed to calculate the expected return time for a flood can be given in terms of the the physical model described in Sec. 2. This calculation requires the solution of both a partial differential equation and an ordinary differential equation. These equations describe how the probability of a flood changes through time.

The recurrence time $T(v)$ for a flood with volume v in the cross section, is given by

$$T(v) = \frac{D(v)}{P(v)} \tag{8.4}$$

where $D(v)$ is the expected duration of a flood exceeding $V = v$, and $P(v)$ is the probability of exceeding v at a random time. This formula is a result of the renewal-reward theorem [7].

We now focus on the calculation of $D(v)$ and $P(v)$.

From the simplifying assumptions above, the state of the river is described by the volume v of water in the cross-section of interest, and the state of the tributaries $s \in \mathcal{L}^K$. The transition probabilities for this state in matrix form is

$$\begin{aligned} \frac{\partial \vec{F}(v;t)}{\partial t} &= \text{rate from water flow} + \text{rate from Markov chain state changes} \\ &= -(\Lambda - q(v)I)\,\frac{\partial \vec{F}}{\partial v} + M^T \vec{F} \end{aligned} \tag{8.5}$$

where $\vec{F}(v;t) = (F_1(v;t), \ldots, F_N(v;t))^T$ is a column vector of functions

$$F_s(v;t) = \text{Prob}\{V \leq v, \text{ tributary state } S(t) = s; \text{ at time t}\} \tag{8.6}$$

Eq. 8.5 has a unique solution whenever the Markov chain M is irreducible and the discharge $q(v)$ is sufficient to prevent permanent flooding. [8]

The probability of the river exceeding a flood level v at a 'random time' is given by

$$P(v) = 1 - \sum_{s=1}^{N} F_s(v),$$

where $F_s(v)$ is the stationary distribution satisfying $\partial \vec{F}/\partial t = 0$. (When no explicit reference to t is given, the stationary solution is implied.) From Eq. 8.5,

$F_s(v)$ can also be determined from

$$(\Lambda - q(v)I)\,\frac{\partial \vec{F}}{\partial v} = M^T \vec{F}, \tag{8.7}$$

a linear, homogeneous, first-order ordinary differential equation.

$D(v)$ is calculated by simulating floods using Eq. 8.5. Numerical methods for calculating $D(v)$, $P(v)$, and $T(v)$ are discussed in Sec. 5.

4 SAMPLE CALCULATION

The above formulation is sufficient to determine how changes in the river system or discharge rate change the return time of an extreme flood. In this section, sample calculations for a simple system are given to illustrate how the return times of extreme floods when the Chezy equation parameters are changed.

Suppose that either it is raining at λ_r m^3/sec over the catchment, or that no rain enters the river. The basal river flow from springs is λ_2. The river input during rain $\lambda_1 = \lambda_r + \lambda_2$ is the sum of the basal flow and rainfall. Rain lasts an exponential amount of time with mean $1/\mu_{12}$, and periods of no rain are exponential with mean $1/\mu_{21}$. (The exponential model is required for the MMFS.) Then

$$\Lambda = \begin{pmatrix} \lambda_1 & 0 \\ 0 & \lambda_2 \end{pmatrix} \quad \text{and} \quad M = \begin{pmatrix} -\mu_{12} & \mu_{12} \\ \mu_{21} & -\mu_{21} \end{pmatrix}$$

The stationary probabilities of rain vs. no rain are $\vec{\pi}_1 = \mu_{21}/(\mu_{12} + \mu_{21})$ and $\vec{\pi}_2 = \mu_{12}/(\mu_{12} + \mu_{21})$, respectively. The average input to the river is $\bar{\lambda} = \frac{\mu_{21}\lambda_1 + \mu_{12}\lambda_2}{\mu_{12}+\mu_{21}}$.

The river levels for the cases of two differing powers in the Chezy equation ($b = 0$ and $b = 1$) are presented. In both cases,

$$\text{Prob}(V \leq v) = \frac{\mu_{21}}{\mu_{12} + \mu_{21}} F_1(v) + \frac{\mu_{12}}{\mu_{12} + \mu_{21}} F_2(v)$$

where $\frac{\mu_{21}}{\mu_{12}+\mu_{21}}$ is the probability of rain at a random time, $F_1(v) = \text{Prob}(V \leq v,$ during rain), and $F_2(v) = \text{Prob}(V \leq v,$ no rain).

Insights follow the calculations.

Case 1: Discharge proportional to volume, $q(v) = cv$.

The appendix shows that the following functions satisfy the differential equation for the probability of not exceeding a given river level at a random time.

$$\begin{aligned} F_1(v) &= \frac{\mu_{21}}{\mu_{12}+\mu_{21}} I\left(\frac{cv-\lambda_2}{\lambda_1-\lambda_2}, \frac{\mu_{21}}{c}+1, \frac{\mu_{12}}{c}\right) \\ F_2(v) &= \frac{\mu_{12}}{\mu_{12}+\mu_{21}} I\left(\frac{cv-\lambda_2}{\lambda_1-\lambda_2}, \frac{\mu_{21}}{c}, \frac{\mu_{12}}{c}+1\right) \end{aligned} \tag{8.8}$$

with $v \in (\lambda_2/c, \lambda_1/c)$ and $I(z,a,b) = \frac{\Gamma(a+b)}{\Gamma(a)\Gamma(b)} \int_{t=0}^{z} t^{a-1}(1-t)^{b-1} dt$ is the incomplete regularized beta function [9].

The expected duration $D(v)$ of a flood of level v or higher with discharge $q(v) = cv$ is

$$D(v) = \frac{\psi(\lambda_1/c) - \psi(v)}{(\lambda_1 - cv)^{\frac{\mu_{12}}{c}} (cv - \lambda_2)^{\frac{\mu_{21}}{c}}} \tag{8.9}$$

where

$$\begin{aligned} \psi(v) &= \frac{(\lambda_1-\lambda_2)^{(\mu_{12}+\mu_{21})/c}}{c} \times \\ & \left[B(\frac{cv-\lambda_2}{\lambda_1-\lambda_2}, \frac{\mu_{21}}{c}+1, \frac{\mu_{12}}{c}) + B(\frac{cv-\lambda_2}{\lambda_1-\lambda_2}, \frac{\mu_{21}}{c}, \frac{\mu_{12}}{c}+1) \right], \end{aligned}$$

where $B(z,a,b) = \int_{t=0}^{z} t^{a-1}(1-t)^{b-1} dt$ is the incomplete beta function.

Case 2: Discharge constant, $q(v) = d$ when $v > 0$. Although no river actually discharges c m^3/sec when there is water and does not discharge when there is none, this special case illustrates how the river levels are affected by changes in the Chezy power law. The assumption $d > \bar{\lambda}$ is necessary so that the river can drain itself.

$$\begin{aligned} F_1(v) &= \frac{\mu_{21}}{\mu_{12}+\mu_{21}} - \frac{\mu_{21}}{\mu_{12}+\mu_{21}} e^{\zeta v} \\ F_2(v) &= \frac{\mu_{12}}{\mu_{12}+\mu_{21}} - \frac{\lambda_1 - d}{d-\lambda_2} \frac{\mu_{21}}{\mu_{12}+\mu_{21}} e^{\zeta v} \end{aligned} \tag{8.10}$$

where $v \geq 0$ and $\zeta = \frac{d(\mu_{12}+\mu_{21}) - \mu_{21}\lambda_1 - \mu_{12}\lambda_2}{(d-\lambda_1)(d-\lambda_2)}$.

The expected duration $D(v)$ of a flood of level v or higher is constant

$$D(v) = \frac{\bar{\lambda} - \lambda_2}{\mu_{21}(d - \bar{\lambda})} \tag{8.11}$$

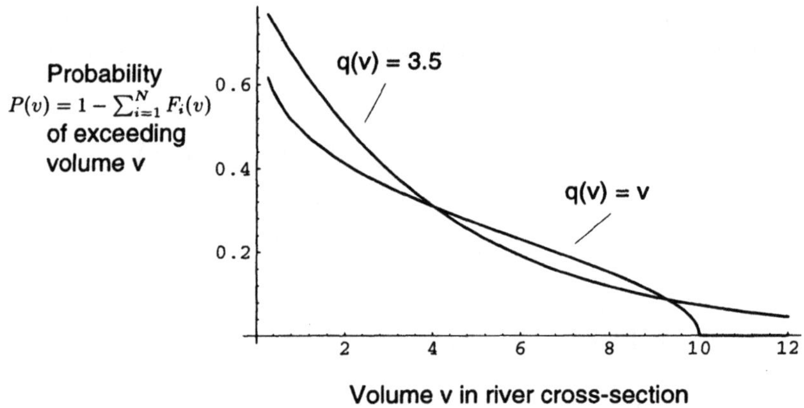

Figure 5 Probability of exceeding a given volume v of water in river cross-section, when $N = 2$, $\lambda_1 = 10$, $\lambda_2 = 0$, $1/\mu_{12} = 2$ days, $1/\mu_{21} = 5$ days.

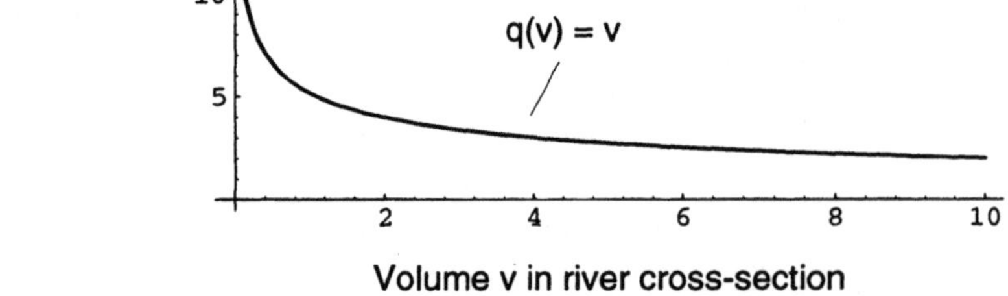

Figure 6 Expected flood duration $D(v)$ same parameters as in Fig. 5.

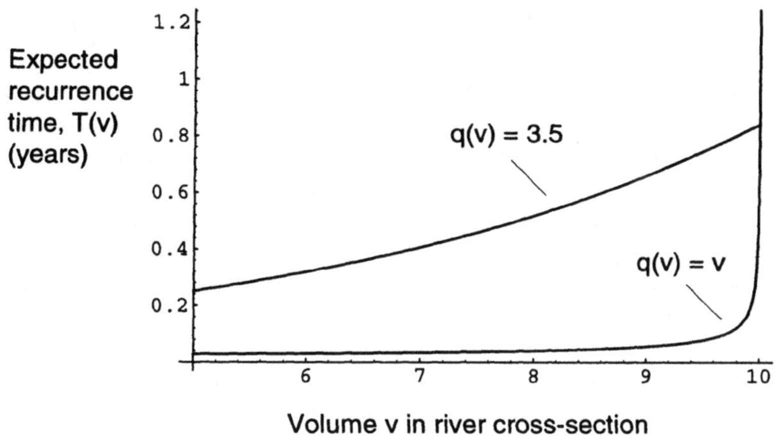

Figure 7 Expected flood recurrence time $T(v)$ using same parameters as in Fig. 5.

whenever $v > 0$. When $v = 0$, the expected flood time is infinity, since the water level is always 0 or greater. The appendix provides proof.

Comments The probability of observing a flood at a random time, the expected flood duration, and the expected flood recurrance time for these two examples are illustrated in Fig. 5, Fig. 6, and Fig. 7, respectively.

An important point is revealed by considering Fig. 7, which relates water levels with the corresponding flood recurrence time. The curves are shaped quite differently because of the change discharge characteristics. If a traditional statistical model has the same curve shape relating flood level and flood frequency, it may accurately reflect the discharge characteristics of the river. On the other hand, we have not yet determined which values of the Chezy parameters, if any, reflect the shapes of traditional statistical models.

Another observation is that increasing the power b from the Chezy equation ($q(v) = av^b$) drastically reduces the recurrence time for a given flood level. The exception is near the asymptote near flood volume $v = 10$, which is near the bounds where the model is expected to blow up. One way to avoid this blow up is have a much larger number of states so that the recurrence time of interest (say, 1 250 years) would not be near the asymptote.

5 GENERAL SOLUTION TECHNIQUES

We turn to the problem of calculating the return times of extreme floods $T(v)$, where v is the volume of water in the river cross-section of interest during a flood. Calculating $T(v)$ (Eq. 8.4) entails the determination of the expected flood duration $D(v)$ and the probabilities $\vec{F}(v)$. The calculation of these quantities requires a solution to Eq. 8.5 and Eq. 8.7. Analytical solutions for extremely simple cases were presented in Sec. 4. More complicated systems, such as for real rivers, seem analytically intractable. A computer based solution for determining $D(v)$ and $\vec{F}(v)$ is required.

To determine $D(v)$, $\vec{F}(v)$ must be calculated first. Two approaches for calculating $\vec{F}(v)$ are available. One approach is to solve the ordinary differential equation (ODE) in Eq. 8.7 as a boundary value problem. Press et al. [10] presents two options for solving ODE boundary value problems, (1) the shooting method, and (2) relaxation methods. We implemented the shooting method for some simple systems, and the algorithm is numerically intensive as well as unstable. The instability is due to the singularities near volumes v such that $\lambda_i - q(v) = 0$. Relaxation methods, which approximate Eq. 8.7 by finite difference equations and attempt to minimize the error of the approximation, seem to overcome this difficulty. We are still in the process of implementing an effective algorithm for large river systems.

When solving with the relaxation method, the following initializations are required. The boundary conditions are $\vec{F}(v_{\text{max}}) = \vec{\pi}^T$ and $\vec{F}(v_{\text{min}}) = 0$, where $v_{\text{max}} = q^{-1}(\lambda_{\text{max}})$, $v_{\text{min}} = q^{-1}(\lambda_{\text{min}})$ are the maximum and minimum possible water levels, and $\vec{\pi}$ is the steady state probability for the Markov chain, with $\vec{\pi}M = 0$.

A second approach for calculating $\vec{F}(v)$ it to integrate the partial differential equation in Eq. 8.5 forward in time until stable values for $\vec{F}(v)$ are achieved. Arbitrary initial conditions will converge to the same stable values, assuming the stability of the numerical methods and the irreducibility of the Markov chain for the state of the tributaries. The ODE relaxation method is preferred, as it is computationally less intensive.

The remaining quantity needed for calculating the return time of floods of level v is the expected flood duration $D(v)$. This may be determined by numerical simulation of Eq. 8.5 through time. The simulation supposes that a flood of volume v has started at time $t = 0$, and keeps track of the probability that the flood recedes below v at each time $t > 0$. From this information, the expected flood duration is calculated as

$$D(v) = \int_{t=0}^{\infty} t \, \text{Prob}(\text{flood recedes at time } t)\, dt \tag{8.12}$$

For the simulation, the flood values and time steps are discretized. This turns Eq. 8.5 into a finite difference equation, and turns the integration in Eq. 8.12 into a sum. No existing algorithms seem ready to solve this equation without some modification. We are presently implementing a simulation based on ideas for PDE numerical methods found in Press et al. [10]

The initial conditions for the simulation use the fact that for a flood of volume v to start, the tributaries must be in a state s such that the rate of water coming in is greater than the rate of discharge, $\lambda_s > q(v)$. The probability of being in state s, then, given that a flood of volume v has just begun, is $F_s(v)/(\sum_{i|\lambda_i > q(v)} F_i(v))$. This leads to the following initial conditions for a simulation to determine $D(v)$.

$$F_s(v'; t=0) = \begin{cases} \frac{F_s(v)}{\sum_{i|\lambda_i > q(v)} F_i(v)} & \lambda_s > q(v), v' > v + \Delta v \\ \frac{(v'-v)F_s(v)}{\Delta v \sum_{i|\lambda_i > q(v)} F_i(v)} & \lambda_s > q(v), v' \in (v, v+\Delta v) \\ 0 & \lambda_s > q(v), v' < v \\ 0 & \lambda_s \leq q(v) \end{cases} \tag{8.13}$$

for some negligibly small $\Delta v > 0$. This simulation calculates $D(v)$ for only a single value of v. For multiple values of $D(v)$, the present formulation requires multiple simulations.

There are asymptotic results for evaluating $D(v)$. Since $D(v)$ is a decreasing function of v, these results may be used to bound the return period for a flood. Suppse the Chezy equation, $q(v) = av^b$, has power $b > 0$, and $\lambda_{\max}$ is the discharge for only one state, $s_{\max}$. Then the expected flood duration is asymptotically the expected time spent in state $s_{\max}$. Formally,

$$\lim_{v \to v_{\max}} D(v) = 1/\mu_{s_{\max}, s_{\max}}.$$

For the case $q(v) = av^0 = a$, the flood duration is asymptotically constant, $\lim_{v\to\infty} D(v) = \text{const}$, where the constant depends on the discharge a and the MMFS, M and Λ.

6 DISCUSSION

The design of flood defense systems requires a statistical model to describe the frequency of extreme floods. There is an increasing awareness that these models should reflect the hydrological information about the river catchments they describe.

Our approach to finding the expected frequency for extreme floods differs in several ways from the more traditional statistical fits to annual maxima (AM) and peaks over threshold (POT) data. For AM and POT data, a line is fit to data points plotted on special paper, such as log paper. These lines have been used to infer the relation between river level and flood return frequency.

In our approach, a continuous time model for the river dynamics is used to determine the frequency of extreme floods. Our sample calculations show the importance of incorporating river dynamics by illustrating the effects of the Chezy equation parameters on the shape of the curves relating the river level and flood return frequency. These shapes do not always conform to the curves found for traditional models. In particular, the relation is not necessarily linear on log paper, as with the Gumbel model.

One benefit of explicitly modeling discharge and other physical factors is the possibility to analyze how changing the river morphology (by nature, dredging, etc.) may affect the severity of extreme floods (100 year, 1 250 year, etc.).

Delft Hydraulics [11] recognizes the importance of modeling the physics of river systems for determining extreme flood frequencies, and has proposed a simulation-based approach to this end. A sophisticated geological information system (GIS) is used to model thousands of datum about a river catchment. Storms of various sizes are simulated, and the high water mark is then calculated. Our approach contains less physics than the Delft Hydraulics proposal, but is a smaller computational burden.

For both our approach and the Delft Hydraulics proposal, there is still a problem with extrapolating what is meant by an extreme storm which might cause an extreme flood. This must be done by specifying a statistical model for the relative strength and direction of movement of storms. While this is a difficult problem, it seems more clear to discuss models of storms than to provide a hydrological basis for the slope of a line on log plots. In addition, there tends to be more meteorological data than AM or POT data. Also, AM and POT data become less usefull as the river morpology changes. Meteorological data is unaffected by structural changes of the river catchment.

7 SUMMARY OF NOTATION

$B(z,a,b)$	incomplete beta function
$D(v)$	expected duration of flood of volume v or greater (time)
δ_i	lag of water flow from tributary i to river cross-section (time)
$F_s(v;t)$	Prob$\{V \leq v,$ tributary state $S(t) = s;$ at time t$\}$
$\vec{F}(v;t)$	vector of $F_i(v;t)$
$\vec{F}(v)$	steady state probabilities for river system
h	height of river (m)
$I(z,a,b)$	incomplete regularized beta function
$q(v)$	discharge from river cross-section (m^3/sec)
$q(v) = av^b$	Chezy's equation for discharge (m^3/sec)
$\bar{q}$	maximal discharge from tributaries (m^3/sec)
K	number of major tributaries which affect river cross-section
$M = \mu_{ij}$	Markov transition matrix for tributary state changes
N	number of states ($\mid \mathcal{L} \mid^K$) of tributaries
$\vec{\pi}$	stationary probability tributary states
l_i	water input rate due to tributary i (m^3/sec)
λ_s	discretized river input when tributaries in state s (m^3/sec)
Λ	diag(λ_s)
$\mathcal{L}$	discrete set of discharge values for tributaries
$S(t)$	state of tributaries, $S(t) = s \in \mathcal{L}^K$ at time t
$T(v)$	expected return time for flood of volume v or greater (time)
V, v	volume of water in a river cross-section (m^3)

8 APPENDIX

Throughout, suppose that $N = 2$, Λ, M, and $\vec{\pi}$ are as specified in Sec. 4.

* Proof of Eq. 8.8. Set $g(v) = \mu_{12}F_1(v) - \mu_{21}F_2(v)$. Eq. 8.7 can be rewritten

$$g' = -\left(\frac{\mu_{12}}{\lambda_1 - cv} + \frac{\mu_{21}}{\lambda_2 - cv}\right) g$$

This has solution

$$g(v) = k_1(\lambda_1 - cv)^{\frac{\mu_{12}}{c}}(cv - \lambda_2)^{\frac{\mu_{21}}{c}}$$

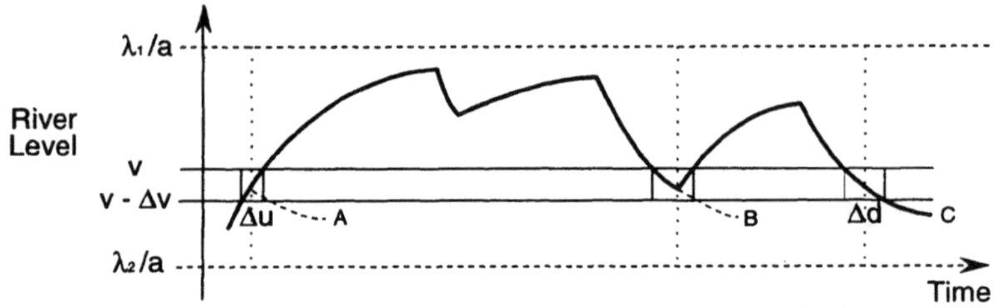

Figure 8 A sample flood of height $(v - \Delta v)$.

where k_1 is some constant. By definition of $g(v)$,

$$F_1'(v) = \frac{-1}{\lambda_1 - cv} g(v) \text{ and } F_2'(v) = \frac{1}{\lambda_2 - cv} g(v)$$

Integrating shows that the F_i are multiples of an incomplete beta function with the parameters given in Eq. 8.8. The boundary condition $F_1(\lambda_1/c) = \pi_1 = \mu_{21}/(\mu_{12} + \mu_{21})$ determines the constant of proportionality.

* Proof of Eq. 8.9

The expected length of a flood $D(v - \Delta v)$ is found in terms of $D(v)$. It is assumed that Δv is small and that $v \in (\lambda_2/a, \lambda_1/a)$. With high probability, the length of a $(v - \Delta v)$ flood is the same as the length of a level v flood *plus* the extra time the river is in the region $[v - \Delta v, v]$ (flood B in Figure 8). These extra up and down crossing times are given by the river flow equation, $\dot{v} = \lambda_i - cv$:

$$\Delta u = \frac{\Delta v}{\lambda_1 - av} + o(\Delta v^2), \qquad \Delta d = \frac{\Delta v}{av - \lambda_2} + o(\Delta v^2). \tag{8.14}$$

It is also possible that the river up-turns in the region $[v - \Delta v, v]$, generating another level v flood (flood C). Another possibility is that the river down-turns before it reaches height v (flood A).

These scenarios are summarized mathematically as follows. The probability that the river does not down-turn in the first time interval Δu is $(1 - \mu_{12}\Delta u) + o(\Delta v^2)$. If there is no down-turn, then there are a series of level v floods, where the number of such floods is a geometric random variable with mean $1/p = (1 + \mu_{21}\Delta d) + o(\Delta v^2)$. The expected length of each such flood is $(\Delta u/2 + D(v) + \Delta d/2) + o(\Delta v^2)$, where the expected extra time spent in $[v - \Delta v, v]$ is included. Finally, the time at the very beginning and end of the flood is added,

$(\Delta u + \Delta d)/2$. This gives the following equation *to first order* in Δv:[2]

$$D(v-\Delta v) = (1-\mu_{12}\Delta u)(1+\mu_{21}\Delta d)\left(D(v) + \frac{\Delta u + \Delta d}{2}\right) + \frac{\Delta u + \Delta d}{2}. \quad (8.15)$$

Substituting Eq. 8.14 into Eq. 8.15 and taking the limit as $\Delta v \to 0$ gives:

$$\frac{dE(v)}{dv} + \left(\frac{\mu_{21}}{av - \lambda_2} - \frac{\mu_{12}}{\lambda_1 - av}\right) E(v) = -\left(\frac{1}{av - \lambda_2} + \frac{1}{\lambda_1 - av}\right). \quad (8.16)$$

Eq. 8.9 is the solution to this first-order linear differential equation (where $\psi(\lambda_1/c)$ is the constant of integration). It can be checked by L'Hopital's rule that Eq. 8.9 satisfies the desired boundary conditions – namely, $D(\lambda_2/a) = \infty$ and $D(\lambda_1/a) = 1/\mu_{12}$.

* Proof of Eq. 8.10 The proof follows the same strategy as for Eq. 8.8. The only difference is that there is no bound on the maximal volume of water, so the boundary conditions are $F_1(\infty) = \mu_{21}/(\mu_{21} + \mu_{12})$, $F_2(\infty) = \mu_{12}/(\mu_{21} + \mu_{12})$. Further, when it is raining it is impossible to have an empty buffer, so $F_1(0) = 0$. These determine all constants of integration.

* Proof of Eq. 8.11 Along similar lines, if the discharge is $q(v) = d$ when $v > 0$, the analog of Equation 8.16 is

$$\frac{dD(v)}{dv} + \left(\frac{\mu_{21}}{d - \lambda_2} - \frac{\mu_{12}}{\lambda_1 - d}\right) D(v) = -\left(\frac{1}{d - \lambda_2} + \frac{1}{\lambda_1 - d}\right). \quad (8.17)$$

The general solution to this first-order linear differential equation with constant coefficients is

$$D(v) = \frac{\frac{1}{d-\lambda_2} + \frac{1}{\lambda_1 - d}}{\frac{\mu_{21}}{d-\lambda_2} - \frac{\mu_{12}}{\lambda_1 - d}} + k_2 \exp\left[\left(\frac{\mu_{12}}{\lambda_1 - d} - \frac{\mu_{21}}{d - \lambda_2}\right) v\right],$$

for some constant k_2. This equation can be simplified to

$$D(v) = \frac{\bar{\lambda} - \lambda_2}{\mu_{21}(d - \bar{\lambda})} + k_2 \exp\left[\left(\frac{\mu_{12}}{\lambda_1 - d} - \frac{\mu_{21}}{d - \lambda_2}\right) v\right],$$

It remains to calculate k_2. The fraction of time that $v = 0$ is $F_2(0) = (d - \bar{\lambda})/(d - \lambda_2)$, by the steady state equations. The expected time the system is

[2] All scenarios which contribute second order terms to this equation are not considered. For instance, the probability of an up-turn *and* a down-turn while the river remains in $[v - \Delta v, v]$ is second order.

empty is $1/\mu_{21}$, so by the renewal/reward theorem

$$F_2(0) = \frac{1/\mu_{21}}{D(0^+) + 1/\mu_{21}}$$

Solving for $D(0^+)$ in terms of $F_2(0)$ and using the steady state solution,

$$D(0^+) = \frac{\bar{\lambda} - \lambda_2}{\mu_{21}(d - \bar{\lambda})}$$

Therefore, $k_2 = 0$.

The implication is that all floods of volume $v > 0$ have the same expected duration. This makes sense, given the memoryless properties of the tributary states and the constant input/discharge rates.

REFERENCES

[1] H. E. J. Berger, *Flow Forecasting for the River Meuse.* PhD thesis, Technical University of Delft, Dept. of Civil Engineering, 1992.

[2] U.S. Water Resources Council, *Guidelines for Determining Flood Flow Frequency.* Reston, VA: Bulletin 17B of the Hydrology Committee, US Government Printing Office, 1982.

[3] J. H. A. Wijbenga *et al.*, "Toetsing uitgangspunten rivierdijkversterkingen." Deelrapport 2, Maatgevende belastingen, Ministerie van Verkeer en Waterstaat, januari 1993. (in Dutch).

[4] P.H.A.J.M. Van Gelder, "De Hoogwaterramp van Limburg in December 1993, een statistische beschouwing." Rijkswaterstaat, Dienst Weg- en Waterbouwkunde, 7 March 1994. (in Dutch).

[5] C. Cunnane, "Review of statistical models for flood frequency estimation," in *Hydrologic Frequency Modeling* (V. P. Singh, ed.), (Dordrecht), pp. 49–95, International Symposium on Flood Frequency Analyses, D. Reidel Pub. Co., 1987.

[6] D. Mitra and A. I. Elwalid, "Statistical multiplexing with loss priorities in rate-based congestion control of high-speed networks," *IEEE Transactions on Communications*, vol. 42, pp. 2989–3002, Nov. 1994.

[7] S. M. Ross, *Stochastic Processes.* New York: John Wiley & Sons, Inc., 1983.

[8] K. Petersen, *Ergodic Theory.* Cambridge: Cambridge University Press, 1983.

[9] S. Wolfram, *Mathematica: A System for Doing Mathematics by Computer.* Redwood City, CA: Addison-Wesley Pub. Co., Inc., 2nd ed., 1991.

[10] W. Press, S. A. Teukolsky, W. T. Vetterling, and B. P. Flannery, *Numerical Recipes in C.* Cambridge: Cambridge University Press, 2nd ed., 1992.

[11] Ministerie van Verkeer en Waterstaat, "Onderzoek Watersnood Maas." Deelrapport 9, Schademodellering, 12 December 1994. (in Dutch).

REFERENCES

9

REVIEW PROBLEM OWNER AND ENGINEERING PERSPECTIVE

Richard Jorissen*, and Matthijs Kok**

**Road and Hydraulic Engineering Division,*
Rijkswaterstaat, Delft, The Netherlands
***Delft Hydraulics, De Voorst, The Netherlands*

1 INTRODUCTION

The Netherlands is situated on the delta of three of Europe's main rivers: the Rhine, the Maas and the Scheldt. As a result of this, the country has been able to develop into an important, densely populated transport nation. On the other hand, large parts of The Netherlands lie below mean sea level and water levels which may occur on the rivers Rhine and Maas. High water levels due to storm surges on the North Sea or due to high discharges of the rivers form a serious threat for the low lying part of The Netherlands. Construction, management and maintenance of flood defences are essential conditions for the population and further development of the country.

Without flood defences much of The Netherlands would be regularly flooded. The influence of the sea would be felt principally in the west. The influence of the waters of the major rivers is of more limited geographic effect. Along the coast, protection against flooding is principally provided by dunes. Where the dunes are absent or too narrow or where sea arms have been closed off, flood defences in the form of sea dikes or storm surge barriers have been constructed. Along the full length of the Rhine and along parts of the river Maas protection against flooding is provided by dikes.

In fact, the low lying part is divided into more than 50 smaller areas, called dike ring areas or polders, which individually are protected by a system of flood defences. The protection offered by flood defences is not absolute. There are no known upper limits to the possible impact of rain and wind. Under extreme weather conditions flood defences may collapse and the low-lying hinterland may be flooded. Even under less extreme conditions the behaviour of flood

R. Cooke et al. (eds.), Engineering Probabilistic Design and Maintenance for Flood Protection, 109–131.

defences is also not fully predictable such that there is always an albeit limited chance of flooding.

The higher, stronger and more reliable the flood defences, the smaller the probability of their failure. Reducing the probability of damage is the benefit side of the consideration of the safety level of flood defences. Reducing the possibility of consequent damage due to flooding is the essential benefit of the level of safety inherent in the flood defences. To provide these benefits, strengthening the flood defences demands major investments from society. This covers not only the money for construction and maintenance. In many case such construction or improvement of the flood defences may affect the countryside, historical villages, natural life or local culture. The demands that are made on the level of protection against high waters therefore also have to be based on a balancing of social costs against the benefits of improved flood defences.

The protection of The Netherlands against flooding will always have the country's full attention. And for different reasons. Firstly, the relevant natural phenomena have a dynamic character and flood defences may thus be affected over time. Sea level rise is the best known example for this, but settlements of the often peaty soil also demand for a more or less regular reconstruction of flood defences. Secondly the balance between costs and benefits can change. For example, the invested capital and the number of residents in threatened areas have strongly grown over the last decades. Currently the total invested capital behind our flood defences is estimated at 4,000 billions of guilders [4]. The balance between costs and benefits can also change as a result of changing social insights, and last but not least, the actual occurrence of floods and flood damage.

2 FLOOD PROTECTION IN THE NETHERLANDS

2.1 History before 1953

The first residents of the current Netherlands delta situated themselves on the so-called high grounds along the rivers or in the high dunes along the Dutch coast. Nevertheless, these high grounds were also regularly flooded. These floods were caused by high river discharges or high water levels on the North Sea. In both cases, these high water levels usually meant that rivers overflowed

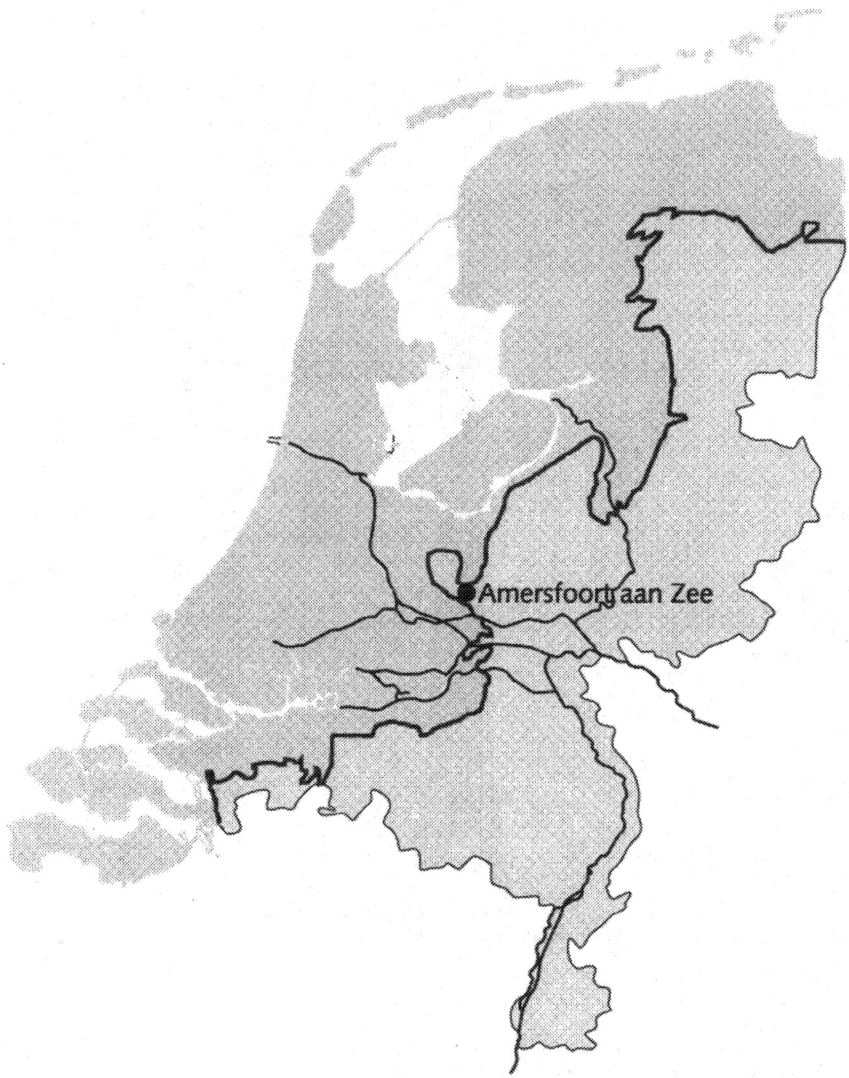

Figure 1 Without flood defences much of The Netherlands would be regularly flooded. Construction, management and maintenance of flood defences are essential conditions for the population and further development of the country. In the low lying parts of the country so-called dike ring areas are formed. Each area is protected by a system of flood defences.

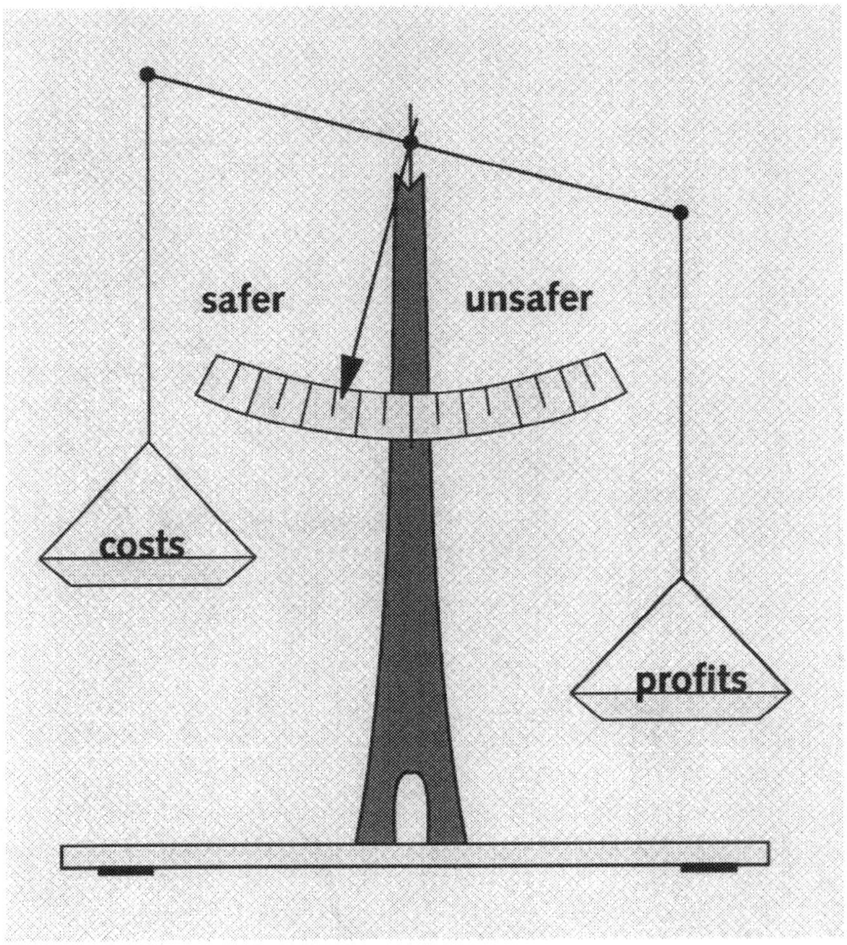

Figure 2 The desired level of safety from protection against flooding means balancing of costs versus benefits of such defence. financial factors; construction and sometimes the reinforcement of existing defences puts demands on scarce land resources. In many case construction or improvement of the flood defences may affect the countryside, historical villages, natural life or local culture.

their banks. Inhabitants of the relatively safe dunes had little to fear. High water levels on the North Sea were to be feared the most, because these caused the most damage and claimed the most victims. Building on elevations, so-called terps or mounds, originates from that early period.

In the centuries that followed, dike construction developed to a national art, in which the use of modern technology, such as steam engines, allowed man to take the offensive. Lakes and sea arms were surrounded with dikes, after which the remaining water was pumped out. The reclaimed areas were extremely fertile. The flood defences of this period were designed primarily based on the highest known water levels.

An important improvement of the protection from flooding was completed in 1933 with the closing of the Zuiderzee. By shortening the coastline, a large portion of The Netherlands was freed from the threats of the sea. Still, The Netherlands remained vulnerable.

2.2 The 1953 disaster

The vulnerability of The Netherlands was made clear in a disastrously painful way in the winter of 1953. A weather depression - combined with an exceptionally high spring tide - caused a storm surge on the North Sea, pushing water levels to record heights. Dikes in the southwestern part of the country failed in several places. In almost all cases, the dikes were too low, causing them to collapse from the 'overflow/overtopping' mechanism.

The direct result of the disaster was 1835 victims and economic damage of NLG 1.5 billion (1956 price index). The indirect economic damage is estimated as a multiple of this. Flooding in Central Holland was barely avoided. The concentration of economic activities in this part of the country would have meant much, much greater damage.

2.3 The Delta Commission

Immediately after the 1953 disaster the Delta Commission was installed. Based on recommendations by this commission, the coastline of South-West Holland was shortened considerably and a more scientific approach for the design of flood defences was implemented. For many areas the shortening of the coastline

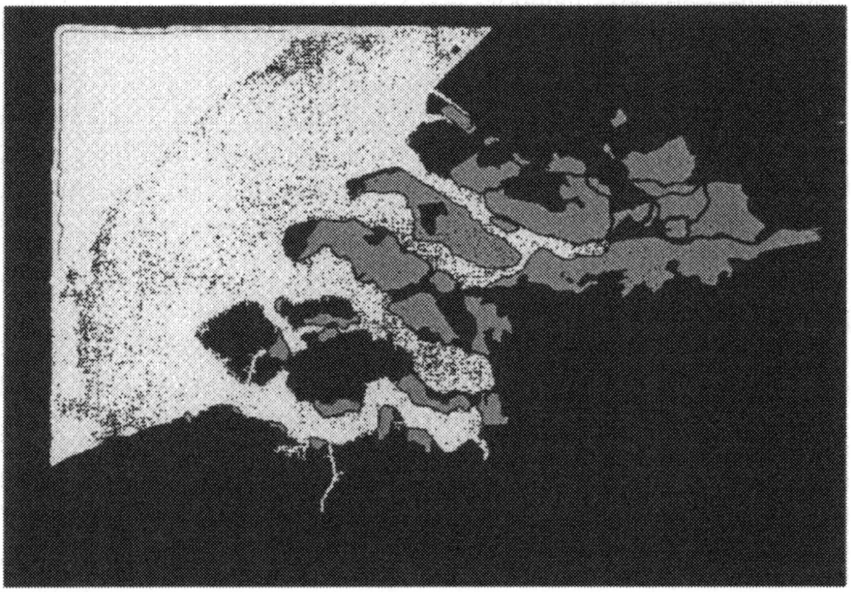

Figure 3 In 1953, the sea dikes failed in many places. In the south west of The Netherlands floods hit on a large scale. Most of the flood defences that failed were simply too low. The overflowing water, sometimes in combination with wave overtopping, caused the initial damage to the dikes' inner slopes. Sliding and erosion of the inner slopes then quickly resulted in dike failures.

by the construction of closure dams largely eliminated the threat of the North Sea.

The design method of flood defences was improved considerably because of the more scientific approach. The standard approach for designing flood defences used until then was based on the highest recorded water level. In relation to this water level a margin of 0,5 to 1,0 metres was maintained. The Delta commission recommended that a certain desired level of safety be taken as a starting point. The safety standards should be based on weighing the costs of the construction of flood defences and the possible damage caused by floods.

An econometric analysis was undertaken by the Delta commission for Central Holland. As far as the economy and the number of inhabitants go, Central Holland is one of the most important dike ring areas. Based on information from 1956 this econometric analysis led to an optimal level of safety, $8 * 10^{-6}$ a year [2] and [3]. This figure represents, in an econometric sense, the optimal required probability of a flood defence failing. This analysis does not include casualties and other imponderables.

Given the available technical capabilities at that time, this safety concept has not been implemented completely. Especially the probability of a flood defence collapsing, and therefore the probability of flooding, proved to be very difficult to estimate accurately. Also, the correlation between the various failure mechanisms has proven an as yet insoluble problem. This is why a simplified safety concept was chosen at the time, based on design loads. The basic assumption with this is that every individual section of a dike has to be high enough to safely withstand a given extreme water level and the associated wave impact. Additional structural requirements, such as the maximum angle of the inner slope and the minimum strength of the cover, assure sufficient stability. The probability of collapse of the flood defence is, however, not determined. This safety concept is therefore referred to as the overload approach.

The design criterium has been based upon the prescribed probability that the extreme water level to be withstood for Central Holland be exceeded. This probability has been set at 1/10,000 per year. This design criterium is based on the enormous consequences of flooding from the North Sea. For economically less important regions, the design water level is based on the acceptance of a higher frequency of flooding. In support of the work of the Delta Commission, the desired safety of the dikes along the rivers Rhine and Maas has been established. Due to the lower potential damage in case of flooding due to high river discharge, the design water level in the upper river area has been fixed at the water level being exceeded with a probability of 1/1250 per year. In this way, a safety level has been established in the form of design water levels for each dike ring area to be protected.

2.4 Current safety policy

Based on the safety concept of the Delta Commission, the national strengthening of the flood defences has been largely completed. The current status is that the coastal barriers (dunes and dikes) have been satisfactorily strengthened. The river dike strengthening is not yet fully complete. Because of the high river discharges of 1993 and 1995, the river dike strengthening will be completed quickly, so that the safety of all flood defences will meet requirements set by the year 2000. Since the founding of the Delta Commission the upgrading of relevant knowledge has progressed strongly. Over the years, the most recent understanding of technically responsible and modern solutions to the flood defence problem have been put into practice continuously.

For dunes, the dike safety concept has been translated into a concept based on the 'dune erosion' mechanism. The basis for this translation process was an equal level of safety for both dune and dike sections. However, the problem that presents itself here is that, in contrast to dikes, the probability of failure of a dune can be calculated with sufficient accuracy. This problem was solved by estimating the probability of failure of a sea dike. Based on the knowledge of the time [5], this resulted in a probability of failure for a sea dike in Central Holland of 1/100,000 per year. Note that this is the probability of failure as opposed to the probability of exceeding the design water level. The current guideline for evaluating the safety of dunes as flood defences is based on this probability of failure [9].

Another expansion concerned the design method of river dikes. The dikes in the lower river area are threatened by both high river discharges and high North Sea levels. In the safety concept for these dikes, both these threats are taken into account. In addition, an important step was taken in the lower river area into the evaluation of the safety of the entire flood defence system. All dike sections form one system for flood protection. Correlation between the different loads and strengths of the various dike sections are taken into account as far as possible. In cases occurring, a conservative approach is taken by assuming independence.

Initially, evaluation of the entire dike ring resulted in a raising of existing requirements. By calibrating the other design criteria, specifically the resistance

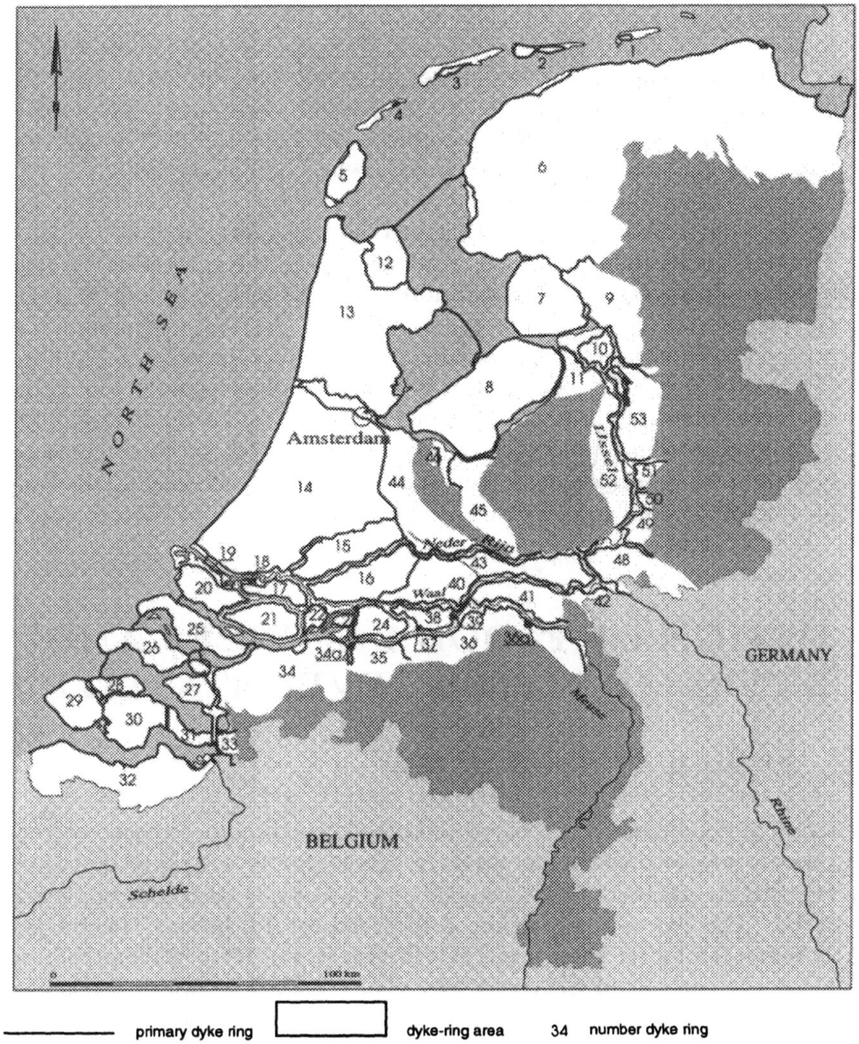

Figure 4 To protect those parts of The Netherlands in which the possible consequences of floods are very serious in terms of number of victims or economic damage, a system of socalled primary flood defences has been constructed. Primary flood defences are defined as those along the North Sea and Waddenzee; and the large rivers the Rhine and Maas; the Westerschelde, the Oosterschelde and the IJsselmeer. The primary flood defences surround a dike ring area to be protected. The Flood Protection Act sets stringent requirements for the primary flood defences. The stringency of the requirements depends on the nature of the flooding and the scope of possible damage. The current safety formulation permits these safety requirements to be expressed in terms of water level that must be withstood. The other, generally more elevated, areas of The Netherlands are naturally safeguarded from large-scale flooding disasters.

of the dike to overtopping and overflowing, the dikes designed are therefore on average as high and as strong as indicated in the recommendations of the Delta Commission. The advantage of this so-called dike ring approach is that the investments in safety can be spent in as profitable a manner as possible. Within certain limits, it is therefore possible in difficult and therefore expensive places to make the dike a little lower and elsewhere a little higher.

From january 1st 1996 for each dike ring area the safety level has been laid down by a new law, the Flood Defence Act. The safety level is a prescribed frequency of exceedance of a water level. Each individual dike section has to withstand this water level and the associated hydraulic loads. No failure is allowed if the design water level occurs. This prescribed frequency of exceedance is uniform along each dike ring area. The criterion varies however between the various dike ring areas, depending on the relative economical importance of each dike ring area. In this way four classes of dike ring areas can be distinguished in terms of water level exceedance frequencies: dike ring areas with a criterion of 1/1250, 1/2000, 1/4000 and 1/10000 per year.

3 DESIGN OF FLOOD DEFENCES AND EXTREME WATER LEVEL STATISTICS

3.1 Design of flood defences

The design of flood defences in The Netherlands is dominated by the hydraulic loads, which may occur. In the old days this hydraulic load was determined in a very simple manner. To the highest kwown water level a safety margin was added. This safety margin ranged from 0,5 tot 1,0 metres. In the years directly before and after the Second World War it was Wemelsfelder [10], who introduced the concept of frequency exceedance curves for water levels along the coast. He drew attention to the fact that the logarithm of the mean number of high tides per year exceeding a certain level resembled a straight line.

The Delta Commission [3] integrated this concept in the design procedure of flood defences. Based on a econometric optimisation the required safety of a flood defence was expressed as a water level, which had to be withstood. As pointed out in the introduction no failure is allowed under the prescribed design conditions. This seems to be a very optimistic demand, because failure of a dike section is always possible. In practice the probability of failure is kept sufficiently low by introducing additional criteria. These deterministic criteria like the steepness of inner and outer slopes of dikes are applied in the practice of dike design. Furthermore there is a minimum margin between crest height and design water level of 0,5 m. And the volume of overtopping water is limited to values ranging from 0,1 to 10 l/m/s.

With regard to the crest height the design procedure is as follows. The required crest height follows from the design water level and a margin for wave run-up or overtopping. The design water level is calculated with a hydraulic model using the design discharge. The margin is determined is such a way, that the probability of exceeding the tolerated overtopping discharges is kept within the prescribed values ranging from 1/10,000 to 1/1,250 per year.

For calculating the required margin not only the design water level is considered. Also combinations of lower water levels with more extreme wind speeds are taken into account. The minimum crest height is 0,5 metres above the design water level, even if there is practically no wind load.

For river dikes the required margin may reach up to 2,0 metres (at locations with long fetches). However in most cases the minimum margin of 0,5 metres or little more is sufficient. This means that the required crest height is almost directly determined by the design water level.

For sea dikes the required margin ranges from 2,5 metres (facing east) to 6,0 metres (facing northwest). So, for sea dikes the design water level doesn't play the crucial role as it does for river dikes. The combined hydraulic load caused by water level and waves has to be taken into account. Both extreme value statistics and correlation between water level and waves are relevant.

3.2 Extreme water level statistics

Although the presented design procedure seems to be fairly simple, there are some hidden pitfalls. Within the topic treated in this paper the main pitfall is how to establish the required frequency exceedance curve of the hydraulics loads. Because of the importance of the water level as the crucial hydraulic load, this question can be narrowed to how to determine the frequency exceedance curve for the discharges ('working' line). The general method applied is as follows.

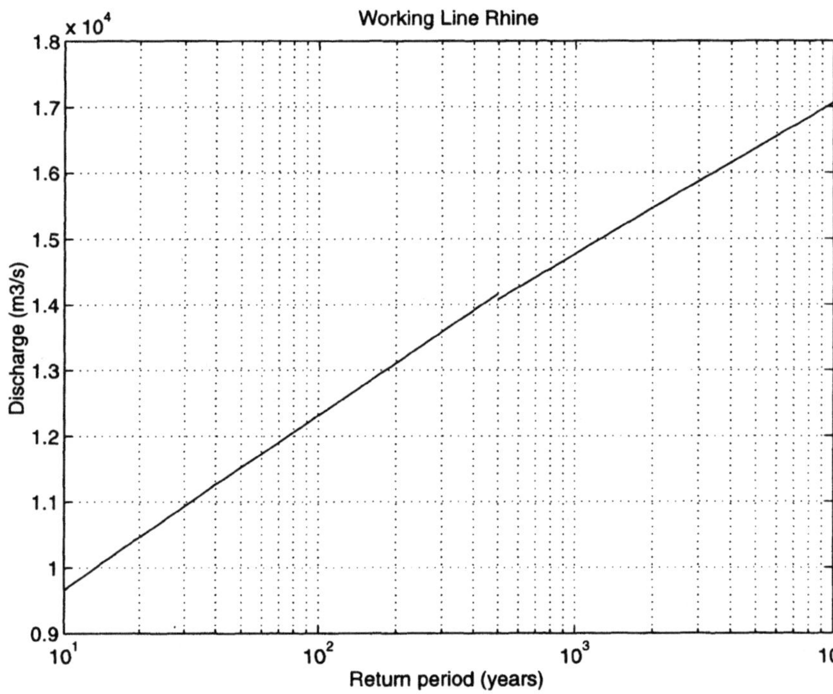

Figure 5 The so-called 'working line' is the applied frequency exceedance curve of the river discharge. This curve shows the discharges of the river Rhine at Lobith, where the river crosses the Dutch-German border. From this 'working line' and the prescribed frequency of exceedance for the river dikes (1/1250 per year) the design discharge can be derived. The design discharge is 15,000 m3/s.

First the available data is collected and analysed. For the river Rhine over a hundred years of continuous water level recordings are available. For the North Sea at Hoek van Holland a similar data set is available. However, not all observations can be used for further analysis. The data set has to be made as homogeneous as possible.

Because of the long period of observations some aspects or elements of the river basin may have been changed. Relevant aspects are changes in ordnance datum, changes in river bed morphology, urbanisation and deforestation, climate changes and altered management of weirs and retention basins. For coastal data the effects of sea level rise have to be taken into account. In the case of historical data the accuracy of the observations is a topic to be dealt with too. for example, the record discharge in the Rhine dates from 1926. The peak discharge however is not exactly known; estimated values range from 12,600 m3/s to 13,500 m3/s. All historical data has to be transformed to hindcasted 'observations' taking into account all relevant changes. In practice, numerical hydraulic models are used for this transformation. Physical scale models may be applied for complex local situations. For the rivers Rhine and Maas these models are also used for the transformation of water levels to river discharges.

After collecting and analyzing the data a selection is made for all those maxima exceeding a certain, high value. Every peak discharge exceeding the threshold is selected for further use. The subset of data above the threshold is analyzed further. To discriminate independent floods the selected peak discharges must satisfy a minimum time-gap. For the river Rhine this minimum time-gap is 15 days. If two consecutive peak discharges occur within a 15-days span, the smallest peak is considered to be dependent on the larger peak. Only the larger peak is considered. For the North Sea the minimum time-gap is not fixed. The Delta Commission considered all potential dangerous depressions following specified tracks and causing a surge effect exceeding 0,5 metres.

After this sequence of collecting, transferring, analysing and selecting a set of independent peak discharges has been obtained. Each peak is considered to be an independent sample from a probability distribution. To determine the distribution the data is plotted and a 'suitable' distribution is compared with the results. Generalized Extreme Value distributions are in most cases suitable distributions. The well-known Gumbel distribution is a special and often used GEV-distribution if the observations are yearly maxima. The GEV is a three-parameter distribution. The parameters are chosen such that the GEV- distribution fits best to the empirical distribution of peak discharges. For the Rhine the maximum likelihood estimation has been used.

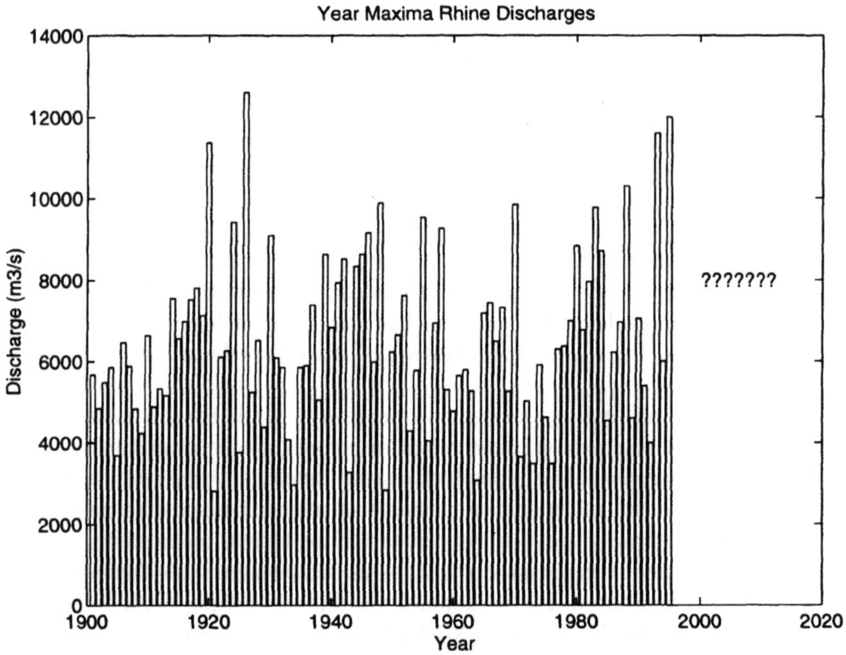

Figure 6 The series of high discharges in the Rhine over the last century shows that the occurrence of high flow rates is unpredictable. This data has been constructed from the available water level recordings. The measured data had to be adapted for changes in ordnance datum, changes in river bed morphology, altered management of weirs and not in the least the transformation from water level recordings into discharges.

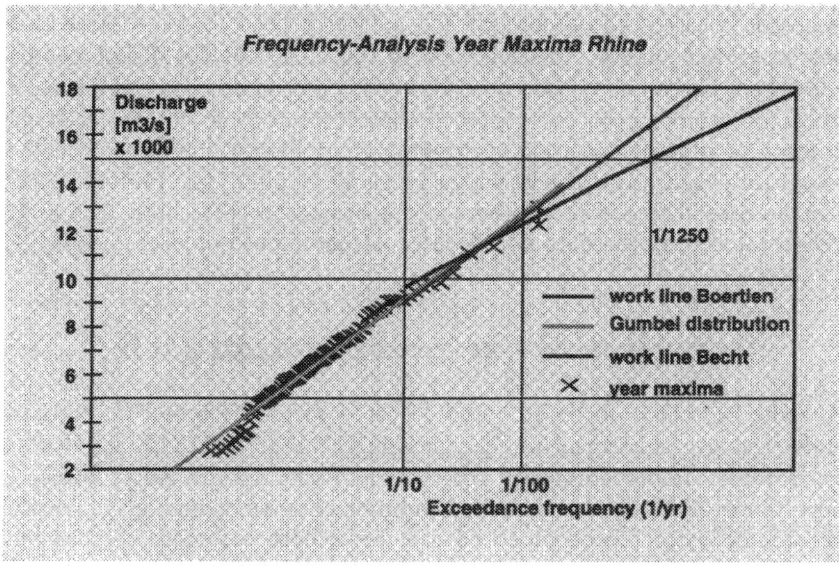

Figure 7 Over the last ten years various GEV-distributions have been applied to the available data for the discharges in the river Rhine at Lobith. From the graph it is obvious that the selection of the suitable GEV-distribution has a significant influence on the design discharge (1/1,250 per year). The design discharges range from 15,000 m3/s to 18,000 m3/s only depending on the selection of the GEV-distribution.

Apart from the preparation of the data the method is a relatively straightforward 'peak over threshold' method. The crucial elements of the method are:

- collecting the available data;
- homogenizing the data;
- selecting the threshold;
- selecting a 'suitable' GEV-distribution;
- fitting the parameters of GEV-distribution.

It is obvious that the selection of both the threshold and the suitable GEV-distribution are subject to many discussions. Especially when the relation between the selected values and the design water level becomes clear to the various interested parties. The selection of the most suitable distribution has to be based on the fundamental knowledge of the mechanism of peak discharges. The tools for acquiring this knowledge are however not available. So, in practice the statistical approach is used to determine the design discharge, which are converted to design water levels with a hydraulic model. The present working line for the river Rhine is determined as the average of three lines. These lines are determined using the Gumbel, Pearson III en lognormal distribution.

3.3 Discussion of the present approach

The present approach of design water levels is primarily based on statistical methods. Using a statistical model a design discharge at Lobith is calculated and a hydraulic model is used to convert this discharge into water levels. Essentially these methods use a limited number of observations (100 years) to predict design water levels, which have a prescribed frequency of exceedance of 1/1,250 per year for the river Rhine. For the North Sea the prescribed frequency of exceedance is 1/10,000 per year. Statistical methods have a large drawback that physical knowledge is not used in this extrapolation exercise. For example, the river basin may react differently under more extreme conditions than that experienced until thus far. Another drawback of the statistical methods is the dependency of recent observations of extreme water levels or discharges. For example, before the flood of 1995 the design discharge for the river Rhine was 15,000 m3/s. The flood of 1995 featured a estimated discharge of 12,060 m3/s. Using this observation and identical statistical methods, the design discharge would increase to 16,050 m3/s. For the upper river areas this would raise the design water levels approximately 0,40 metres.

The lack of physical knowledge is in our opinion the key problem. Other minor problems applying the present POT- method are related to or caused by this key problem. These minor problems are :

- which observations to include in the analysis ?
- which GEV-distribution to apply ?

In practice these minor problems cause a lot of discussion among technicians and between technicians and decision makers. An important item of these discussions is the function of design water levels. In a technical, statistical sense the design water level is just the water level with a frequency of exceedance of 1/1,250 per year. Given the limited number of observations and the unknown behaviour of the river basin during extreme discharges, the design water level will be uncertain in a statistical sense. For the present design water levels on the river Rhine the estimated standard deviation of the design water level is about 0,60 metres, [8]. On the other hand, the design water level is for decision makers their measure of safety against flooding. Design water levels are therefore officially declared by the Minister of Transport, Public Works and Water Management. The present reinforcement program of the river dikes is based on these official design water levels.

It is obvious that these two interpretations of design water level must lead to discussions. Discussions, which can only be solved if the definition and application of design water levels are clarified. The problem of definition and application of design water level can be solved only if statistical uncertainties of design water levels are included in the design procedure for river dikes. Every technical improvement of the present, statistical method will contribute to the solution by reducing the uncertainties, but will not solve the problem as a whole.

4 DISCUSSION OF THE PROPOSED APPROACH BY S. CHICK

4.1 The proposed approach

[1] proposes an approach which introduces river dynamics. By modelling the inflow of water (via upstream tributaries) and the outflow of water (discharge modelled by Chezy equation) at a given location the model accounts for river dynamics.

For each relevant cross-section of a river the major tributaries are identified. The state of the various tributaries are modelled as Markov modulated fluid sources. The basic components of this model are state and fluid rate. The state is defined as the array of discharges from the tributaries. The fluid rate is the sum of these discharges and equals the input of water in the cross- section.

The model supposes the following simplifications. The discharges from the tributaries have to form a discrete set. By prescribing a sufficiently large set this simplification will not cause any problems. The transitions between the various states are memoryless. The catchment is modelled as a continuous-time Markov chain. In practice, this means that an increase of flow has to follow the given discrete set of discharges. On first sight this is not very likely to cause any problems either. But there is a potential pitfall here. On the one hand, one may need a rather large set of discrete discharges. On the other hand each discharge must be 'passed' during the transitions. These two demands may cause the time-step of the model to be relatively small.

Using the available data from the Rhine and its main (9) tributaries, the model can be validated for the past century. This calibrated model can be run as a continuous time model to determine the frequency of extreme floods. Following this procedure a frequency exceedance curve can be constructed without having to select a GEV distribution. The main elements of the proposed method are now :

- collecting the available data;
- homogenizing the data;
- simulating all relevant events, including the extreme events.

4.2 Discussion

The proposed method introduces a limited quantity of physics into the statistical approach. The rainfall- discharge relation for the various tributaries can be introduced into the model. A simplified discharge relation completes the model. The model of Chick enhances the possibilities of predicting the extreme discharges in the catchment area. A very attractive element of the proposed model is the limited computational burden.

Another advantage of the model of Chick is the increased insight in the behaviour of the tributaries. By splitting the entire river system into smaller catchments, it may become easier to describe the (extreme) events in that catchments. If the behaviour of the various catchments can be determined accurately enough from available data, the proposed method will certainly be an improvement to the present method. For example, the flood of 1995 in the Rhine can be distributed over nine tributaries. Of these tributaries only one reached its highest known value. The total discharge of the Rhine in 1995 was 12,060 m3/s. In 1993 a peak discharge of 11,600 m3/s occurred in the Rhine.

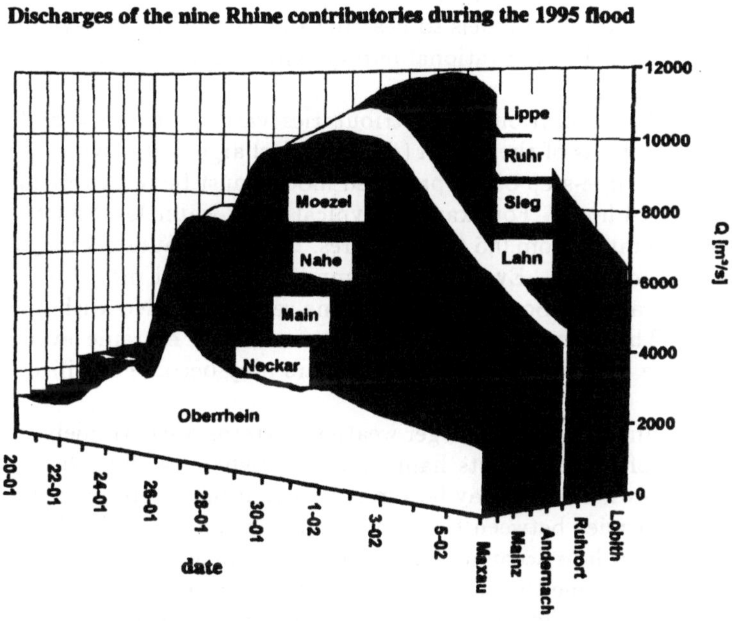

Figure 8 The behaviour of the tributaries is the basis for the entire catchment area. The flood of 1995 in the Rhine can be distributed over nine tributaries. Of this tributaries only one reached its highest known value. The total discharge of the Rhine in 1995 was 12,060 m3/s. In 1993 a peak discharge of 11,600 m3/s occurred in the Rhine. During this lower peak discharge period three minor tributaries reached their highest known values. Chick's model may give more insight in how the tributaries contribute to the behaviour of the entire catchment.

During this lower peak discharge period three minor tributaries reached their highest known values.

Although the advantages of the model are clear, a few questions remain.

The larger river basins form a very complex system. The behaviour of such systems is only partially described by the proposed discharge relation in the model of Chick. Flood plains and retention basins influence the behaviour of the river basin during extreme events. Human interference and future policy regarding the use of flood plains may alter. Consequently, discharge relations may change in time, even during flood periods. These events are not included in the proposed model. It seems to be possible to introduce available but more complex hydraulic models to describe the catchment behaviour, but the advantage of limited computational burden will be given away then.

The 'typical' flood duration of the tributaries varies between 3 and 10 days. Given the limitations of the model (discrete discharge states which all have to be 'passed') the time-step of the proposed model must be relatively limited, for example 12 to 24 hours. For example, typical daily water level fluctuations on the river Rhine range from 0,5 to 1,0 metres. The contributaries may even vary on a smaller time scale. Even for the simplified physical model no analytical solutions are available. So simulation of time-series is needed with many states to consider. The computational burden to calculate extreme events with a recurrence time between 1,000 en 10,000 years may become significant.

For larger catchment areas the larger weather patterns are extremely important. The exclusion of certain events happening somewhere, while other events are happening somewhere else, may become important in determining the extreme events. Dependencies between the various large- scale events in the catchment area are not taken into account. For example, the weather pattern causing the snow in the Alps to melt in winter is in many causes also causing a lot of rain in Europe. These coinciding events may cause extreme discharges.

The need for improving the models for predicting extreme water levels is evident. Extreme water levels play a significant role in designing Dutch flood defences. The present statistical methods are insufficient to predict the behaviour of the catchment area during extreme events. The model of Chick is an improvement of the present statistical models, however the key problem has not been solved yet.

4.3 Alternative proposal

The lack of physical knowledge and/or information is in our opinion the key problem. This key problem can only be tackled by developing a combined hydrological/hydraulic model of the entire catchment area. The attractiveness of this alternative is that all components are readily available. Each section of the Rhine and its catchment area is somehow modelled. Integration of these models into one model has proven to be impossible sofar due to incompatible models and systems, computational burden and available data.

By simulating all relevant events with a combined hydrological/hydraulic model all relevant physical events can be described. Even future policies regarding the use of flood retention basins can be integrated. The problem is how to supply the model with the boundary conditions and the computational burden of this approach.

The required boundary conditions must be selected carefully. It is of no use to simulate all known events. It is relevant to restrict the simulation to the events of interest. To determine these events a study into the causes of peak discharges is necessary. If the relevant large scale meteorological boundary conditions can be described with a limited number of (statistical) parameters the computational effort may be limited. The combined hydrological/hydraulic approach also benefit from the fact that the natural correlations are included 'automatically'. The simulation method has been applied recently for designing the storm surge barrier near Rotterdam [6] and [7].

The computational burden is to our opinion a lesser problem. Partly it will be solved by the growing capacities of modern computers. The remaining effort can be resolved by carefully selecting the boundary conditions to be simulated and by optimizing the use of the models. Generally, this will lead to a minimisation of the on-line use of mathematical models and maximising the use of earlier calculated data. One could introduce so-called 'reproduction' functions to determine the discharges at various locations based on input from other models. For the storm surge barrier near Rotterdam [6] and [7] the physical calculations have been completely isolated from the statistical calculations.

5 CONCLUSIONS AND RECOMMENDATIONS

In The Netherlands flood defences provide safety against flooding for a majority of the country, both in terms of surface area, inhabitants and economical importance. The safety level of the dutch flood defences is presently expressed as the frequency of exceedance of water levels. In combination with wind generated waves the water level is the most dominant hydraulic load.

Especially along the rivers Rhine and Maas the required crest height of the dikes is almost directly determined by the design water level. The design water level for Dutch flood defences has a prescribed frequency of exceedance ranging from 1/1,1250 per year to 1/10,000 per year.

The present method of determining the design water level is the 'peak over threshold' method. The main element in using this method are :

- creating a sufficiently large and sufficiently homogeneous data set;
- selecting a 'suitable' GEV-distribution;
- extrapolating the fitted GEV-distribution far outside the area of observations.

The present statistical method is used because of the absence of tools for a more physical approach. This lack of physical knowledge is the key problem of the present method.

Chick proposes a model, which is still a statistical model. The proposed model of Chick has the advantage of encapsulating some physical knowledge about river dynamics. For studying the relative importance of the various tributaries the proposed method seems to offer very good opportunities. The present POT-method doesn't include any physical knowledge at all. The physics introduced in the proposed model however is only sufficient for the known situations and the situations which may be constructed using the known behaviour of tributaries. For extreme events the introduced physics is not enough. The proposed model doesn't facilitate the study into future human interference.

As an alternative the combined hydrological/hydraulic model for the entire catchment area is proposed. Such a model contains all the physics required. If boundary conditions are derived carefully and computational effort is minimized the computational burden of a combined hydrological/hydraulic model may be in the same order of magnitude as the model proposed by Chick.

REFERENCES

[1] Shortle J. Gelder P.H.A.J.M. van Mendel M. Chick, S. A physics-based approach to predicting the frequency of extreme river levels. 1995.

[2] D. van Danzig. Economic decision problems for flood prevention. *Econometrica*, 24:276–287, 1956.

[3] Rapport Deltacommissie, Deel 3, Appendix II.2, Het economisch beslissingsprobleem inzake de beveiliging van Nederland tegen stormvloeden

[Report of the Delta Commission, Part 3, Appendix II.2, The economic decision problem of flood defence in the Netherlands]. The Hague, The Netherlands, 1960.

[4] DGW (National Institute for Coastal and Maritime Management). Analysis of vulnerability to the impacts of sea level rise. *Resource Analysis*, DGW-93.034:917–931, 1992.

[5] J. van der Graaff. Probabilistic design of dunes; an example from the netherlands. *Coastal Engineering*, 9:479–500, 1986.

[6] J.P.F.M. Janssen and Jorissen R.E. Modelling storm surges and forecast effects in designing the rotterdam storm surge barrier. 1991.

[7] J.P.F.M. Janssen and Jorissen R.E. Integrating forecast effects and operational behaviour in designing the rotterdam storm surge barrier. 1992.

[8] Douben K.J., van Noortwijk J.M., Silva W, and Kok M. Landscaping river Rhine : Safety (in dutch). 12, 1995.

[9] TAW (Technical Advisory Committee on Water Retaining structures), CUR (Centre for Civil Engineering Research , and Codes). Guideline to the assessment of the safety of dunes as a sea defence. 1989.

[10] P.J. Wemelsfelder. Wetmatigheden in het optreden van stormvloeden [Laws of nature in the occurrence of storm surges]. *De Ingenieur*, 54(9):B.31–35, 1939.

10

REVIEW MATHEMATICAL PERSPECTIVE

Jolanta Misiewicz

Department of Mathematics,
Technical University of Wrocław, Wrocław, Poland

The authors proposed a new method for predicting the frequency of extreme river levels, a method based on physical knowledge of river dynamics including the Chezy equation for discharge and water input from tributaries.

The rate of discharge $q(v)$ from a cross-section of a river is a function of the geometry of the cross-section and the volume of water v in that cross-section, and it is described by Chezy formula

$$q(v) = av^b.$$

As I understand, the constants a and b have to be approximated statistically. On the other hand we have K major tributaries transporting water into the cross-section, and the flow time from tributary i to the river cross-section is approximately δ_i. Then the rate of water input $\lambda(t)$ is given by

$$\lambda(t) = \sum_{k=1}^{K} l_i(t - \delta_i),$$

where $l_i(t)$ is the discharge in m^3/sec from tributary i at time t. Again, the functions l_i have to be approximated statistically.

The authors model the state of tributaries as a modulated, continuous time Markov chain assuming a finite number of possible states $\mathcal{L} = \{0, 25, 50, \ldots\}$ for tributaries, so that the river catchment at time t on the river cross-section is described by $S(t) = (l_1(t - \delta_1), \ldots, l_K(t - \delta_K)) \in \mathcal{L}^K$. As a continuous time Markov chain process $S(t)$ is memoryless with the transition matrix $M = (\mu_{ij})$. Although M is a large matrix, the authors assume that $\mu_{ij} = 0$ whenever $|i - j| > 2$ which is a reasonable assumption and simplifies calculations. It is

R. Cooke et al. (eds.), Engineering Probabilistic Design and Maintenance for Flood Protection, 133-134.

hard to find in the paper, but the authors do not assume independence between tributaries, so that tributaries coming from the same area should be correlated.

These are the assumptions and all the results in the paper are calculated on this continuous time Markov chain construction. Evidently the proposed method of investigations is new for flood frequency and water high level crossing, where most often extremal statistics are in use. However I do not think that the authors justified this method carefully enough. In particular, the paper should contain answers for the following two questions:

- Does the assumption that the stochastic process of states of tributaries $S(t)$ is memoryless approximate the real situation sufficiently well? Somehow it is easier for me to believe that tomorrow will be raining if it was raining for the whole last week, than in the situation that it was raining only today, but not yesterday. The answer for this question should probably be supported by some statistical analysis.
- What is the influence of small errors in approximation of the functions $l_j(t)$ and approximation of elements of transition matrix $M = (\mu_{ij})$ to the final result of calculations? The number of approximated functions and coefficients is so large that all the errors could cumulate easily.

Finally, I have found the paper mathematically correct and interesting. For real applications however much more work should be done in justifying these nice ideas.

11

REPLY TO THE REVIEWS

Steve Chick,
John Shortle*, Pieter van Gelder** and Max Mendel*

Department of Industrial and Operations Engineering,
University of Michigan, Ann Arbor, Michigan 48109
**Department of Industrial Engineering and Operations Research,*
University of California, Berkeley, California 94720
***Department of Civil Engineering,*
Delft University of Technology, Delft, The Netherlands

1 SUMMARY RESPONSE TO CRITICISMS

The commentaries of Jorissen and Kok, and Misiewicz are thorough. Each admits that the proposed MMFS/discharge model provides a number of advantages. Among them is the encapsulation of hydraulic equations for river discharge into a statistical model which predicts return times for extreme floods. Further, the model includes the effects of multiple tributaries to account for their joint contribution to flooding.

At the same time, a number of questions and criticisms remain. While the proposed MMFS/discharge model encapsulates much more physics of river dynamics than traditional statistical models, it contains an enormous number of parameters. This large number of parameters puts significant demands on data collection (to 'tighten' posterior distributions for those parameters) and on computation. The trade-off between the beneficial addition of some physics with the negative effect of a large number of parameters is unclear.

Additionally, the proposed MMFS/discharge model contains much less physical information than an extensive hydrological/hydraulic model. Clearly, the MMFS/discharge model does not contain the level of detail of a sophisticated model containing, for example, large quantities of detailed geographic and river morphology data. Further, the problem of trade-offs between a small discretization between discharge rates for the MMFS and the data collection requirements to support the effective analysis of MMFS parameters are very real. Ideally,

R. Cooke et al. (eds.), Engineering Probabilistic Design and Maintenance for Flood Protection, 135-138.

one would like a significant amount of data to determine 'narrow' posterior distributions for these parameters.

In the face of these criticisms, what is the usefullness of the MMFS/discharge model? We return to the original motivation for the MMFS/discharge model to respond to this question. The original goal was to provide intuition for how changes in river dynamics affect the return period for floods. While this intuition may be somewhat hidden by the large number of parameters in the proposed model, one can obtain this intuition by considering simplified versions of the MMFS/discharge model. One such simplified version proposes a Chezy law (with two parameters) for the discharge; a MMFS with states $\{0, 1, 2, \ldots\}$; a linear, state-dependent discharge, $\lambda_s = s$; and a rate matrix corresponding to a single tributary, and a probability of increasing of p, with $0 < p < 1/2$, so that $M = (\mu_{ij})$ has

$$\mu_{ij} = \begin{cases} -p & \text{if } i = j = 1 \\ p - 1 & \text{if } 1 < i = j < N \\ 1 - p & \text{if } i = j + 1 \\ p & \text{if } i = j - 1 \\ 0 & \text{otherwise} \end{cases}$$

Such a simplification has advantages over both POT models as well as hydrological/hydraulic models. It is attractive in that it has only three parameters (two Chezy parameters and p), just like many statistical models currently in use, such as the generalized extreme-value model. It is further attractive in that affects of changes in hydraulic conditions (e.g. Chezy parameters) can be modeled directly, even without the extensive geological and meteorological data that a hydrologic/hydraulic model might provide. Thus, for river-basins which are not known as thoroughly as the Maas, Rhine, and Waal, some intuitions may be gained about the affect of the Chezy parameters on the return period for extreme floods. While not pretending to be as sophisticated as an extensive hydrologic/hydraulic model, the proposed model can still provide basic insights into the fundamental problem. Namely, how do values of the Chezy parameters determine the 'curvature' for a 'working-line' relating flood volume and recurrence frequency.

An example of such an insight can be borrowed from communications. The result says that for an arbitrary MMFS (finite number of states) with constant discharge $q(v) = a$, the expected return time for extreme floods is asymptotically exponential. In other words, for constant discharge, the 'working line' for floods should linearly relate the flood volume and the log of the return time. Curiously, the frequency exceedence curves of Wemelsfelder mentioned by Jorissen and Kok follow this same relation.

But what if there is an increase in the exponent b of the power law $q(v) = av^b$? Intuitively, as the exponent of the power law increases, drainage is better and the return time for extreme floods should increase (namely, $\partial T/\partial b \leq 0$). Similarly, one can prove that return periods decrease for increasing volumes, $\partial T/\partial v \leq 0$. An important remaining question is how the slope of the curve relating return periods T with flood volume v changes with b, $\partial^2 T/\partial v \partial b$. If this quantity is positive for large v, then the 'working line' will tend to curve downwards. It is an open question whether this can be proven analytically to be true or not.

Other interesting questions relating to this research relate to the difficulty of obtaining numerical solutions to the differential equations. This is due to a number of factors, including singularities in the differential equations, large order of magnitude differences between the various values of exceedence probabilities for different Markov chain states at a given volume, and other considerations.

2 COMMENTS ON HYDROLOGIC/HYDRAULIC MODELS

The hydrologic/hydraulic model discussed by Jorissen and Kok presents a number of advantages, given that one has sufficient meteorological and geological data to describe the river system and its inputs. One problem the hydrologic/hydraulic model seems to share with the MMFS/discharge model is that the probability of weather patterns needs to be modeled. (Such a model must be specified explicity for the hydrologic/hydraulic model, and is specified implicitly via the Markov matrix M for the MMFS/discharge model.)

Supposing that such a model were available. For the MMFS/discharge model, no simulations are necessary—solutions to differential equations give the desired solution. For the hydrologic/hydraulic model, simulations are required to estimate the volume of water present for a flood with a given return period. The most simple simulation approach, i.e. sampling from the weather patterns and calculating flood volumes, will require an extremely large number of replications (tens of thousands for the flood recurrence times of interest). At the same time, many fields have used variance reduction techniques such as importance sampling to achieve orders of magnitude reductions in the number of replications required. The application of these variance reduction techniques for the special case of flood prediction in Holland therefore appears to be an interesting and valuable line of research.

PART III

12

LARGE MEAN OUT CROSSING OF NONLINEAR RESPONSE TO STOCHASTIC INPUT

Chun-Ching Li, and Armen Der Kiureghian

Department of Civil Engineering,
University of California, Berkeley, California 94720

ABSTRACT

The mean out-crossing rate of a nonlinear system subjected to stochastic excitation is computed by use of the first-order reliability method, FORM. The mean out-crossing rate can be used to compute an upper bound to the failure probability. The method can account for system stochasticity and can be extended to multi-degree-of-freedom systems. Three systems are investigated: a linear oscillator, a duffing oscillator, and a hysteretic oscillator.

1 INTRODUCTION

In structural reliability, the probability that the response of a structure subjected to stochastic loading exits a specified safe domain during a time interval $0 \leq t \leq T$ is of great interest. For example, the probability that story drifts in a multi-story building subjected to stochastic earthquake excitation out-cross a domain defined by a set of allowable thresholds might be used as a measure of safety. Most generally, this problem is formulated in terms of one or more limit-state functions $G(\mathbf{u(t)}, \mathbf{v})$, expressed in terms of a vector of response processes $\mathbf{u(t)}$ and a vector of system variables $\mathbf{v}$. Included in $\mathbf{v}$ are parameters defining the inertia and constitutive properties of the system (e.g., mass, stiffness, damping, inelastic material properties), which may be considered random to account for system stochasticity. The vector $\mathbf{u(t)}$, representing such responses as nodal displacements, stresses or internal forces, is dependent on $\mathbf{v}$, the vector of input processes, $\mathbf{F(t)}$, and the initial conditions of the system. Its evolution in time is governed by a system of differential equations describing the motion

R. Cooke et al. (eds.), Engineering Probabilistic Design and Maintenance for Flood Protection, 141–160.

of the system, and the constitutive laws of the materials.
By convention, the function $G(\mathbf{u}, \mathbf{v})$ is defined such that it assumes a positive value in the safe domain, negative value in the unsafe domain, and zero on the boundary between the two domains. Hence, the probability of failure over a duration $(0, T)$ can be written as

$$P_f(T) = P(\{\min_{0 \leq t \leq T} G(\mathbf{u}(\mathbf{t}), \mathbf{v})\} \leq 0) \tag{12.1}$$

Another way to define the failure probability is in terms of out-crossing events. An out-crossing event occurs when the tip of the vector $\mathbf{u}(\mathbf{t})$ lies on the surface $G(\mathbf{u}, \mathbf{v}) = \mathbf{0}$ and its derivative $\dot{\mathbf{u}}(t)$ has a positive projection along the outward normal to the surface. The probability of failure is the complement of the event that $\mathbf{u}(\mathbf{t})$ starts within the safe domain and it has zero out-crossings during the interval $0 < t \leq T$, i.e.,

$$P_f(T) = 1 - P\left(G(\mathbf{u}(\mathbf{0}), \mathbf{v}) > \mathbf{0} \cap \mathbf{N}(\mathbf{T}) = \mathbf{0}\right) \tag{12.2}$$

where $N(T)$ is the number of out-crossings in $(0, T)$, a discrete random variable. For structural reliability problems, the event $G(\mathbf{u}(\mathbf{0}), \mathbf{v}) > \mathbf{0}$ usually has almost 1 probability; that is, the system normally starts from within the safe domain. Hence, we may write (Lin [6]),

$$\begin{aligned} P_f(T) &\approx 1 - P(N(T) = 0) \\ &= \sum_{n=1}^{\infty} P(N(T) = n) \\ &\leq \sum_{n=1}^{\infty} nP(N(T) = n) \\ &= \mathrm{E}[N(T)] \end{aligned} \tag{12.3}$$

where $\mathrm{E}[\cdot]$ denotes the expectation. An upper bound to the failure probability, thus, is

$$P_f(T) \leq \mathrm{E}[N(T)] = \int_0^T \nu(t)\, dt \tag{12.4}$$

where $\nu(t)$ is the mean rate of out-crossing per unit time. This upper bound works well when $P_f(T)$ is small and the likelihood of two or more out-crossings is negligible compared to that of only one out-crossing.

It is apparent from the above analysis that the mean out-crossing rate $\nu(t)$ plays an important role in determining the failure probability of systems under stochastic loading. This rate may be computed by applying Rice's formula (Rice

[12], [13]) to the zero-level down-crossings of the scalar process $G(\mathbf{u(t)}, \mathbf{v})$,

$$\nu(t) = \int_{-\infty}^{0} |\dot{G}|\, f(G = 0, \dot{G})\, d\dot{G} \tag{12.5}$$

in which $f(G, \dot{G})$ is the joint probability density of $\dot{G}(\mathbf{u(t)}, \mathbf{v}) = dG(\mathbf{u(t)}, \mathbf{v})/\mathbf{dt}$ and $G(\mathbf{u(t)}, \mathbf{v})$. This formulation, however, is difficult to compute, primarily because the joint density of G and $\dot{G}$ is difficult to determine for a general $G(\cdot, \cdot)$ function or a nonlinear system.

An alternative is to compute the mean out-crossing rate from the limit formula (Hagen et al [4]),

$$\nu(t) = \lim_{\delta t \to 0} \frac{P(g_1 < 0 \cap g_2 < 0)}{\delta t} \tag{12.6}$$

where $g_1 = -G(\mathbf{u(t)}, \mathbf{v})$ and $g_2 = G(\mathbf{u(t + \delta t)}, \mathbf{v})$. The numerator in the above expression describes the condition under which $\mathbf{u(t)}$ is inside the safe domain at time t and outside the safe domain at time $t + \delta t$. Assuming that for a sufficiently small δt at most one out-crossing event is possible, the above ratio approaches the mean out-crossing rate in the limit as $\delta t \to 0$. The alternative method for computing $\nu(t)$ is to determine the intersection probability in the numerator for a finite but small δt and divide it by δt. The first-order reliability method, FORM (Madsen et al [9]), is well suited for this purpose. The following section describes this approach in more detail.

2 FORM APPROXIMATION

Consider the probability integral

$$P = \int_F f(\mathbf{x})\, \mathbf{dx} \tag{12.7}$$

where $\mathbf{x}$ denotes a set of random variables with joint probability density function $f(\mathbf{x})$, and F denotes a subset of the outcome space. We consider the case where F is described in terms of the intersection of two events,

$$F \equiv \{g_1(\mathbf{x}) \leq \mathbf{0} \cap \mathbf{g_2}(\mathbf{x}) \leq \mathbf{0}\} \tag{12.8}$$

in which $g_1(\mathbf{x})$ and $g_2(\mathbf{x})$ are two limit-state functions. A FORM approximation of the probability P is obtained by the following means:

1. The variables $\mathbf{x}$ are transformed to standard normal variates $\mathbf{y}$ through a transformation $\mathbf{y} = \mathbf{y}(\mathbf{x})$ such that $\mathbf{y}$ has the probability density function $\phi_n(\mathbf{y}) = (2\pi)^{-n/2} \exp(-\frac{1}{2}\mathbf{y}^t\mathbf{y})$, in which n denotes the number of random variables and the superposed t denotes the transpose of a matrix or vector.
2. The surfaces $g_1(\mathbf{x}(\mathbf{y})) = 0$ and $g_2(\mathbf{x}(\mathbf{y})) = 0$ are linearized at points $\mathbf{y}_1^*$ and $\mathbf{y}_2^*$ in the standard normal space. The choice of these points is discussed below. The linearized functions, scaled by the norm of the corresponding gradients, are

$$\frac{1}{||\nabla g_i||} g_i(\mathbf{y})_{\text{linearized}} = \frac{1}{||\nabla \mathbf{g_i}||} \mathbf{g_i}(\mathbf{y}_i^*) + \frac{\nabla \mathbf{g_i}^t}{||\nabla \mathbf{g_i}||} (\mathbf{y} - \mathbf{y}_i^*) \quad \mathbf{i} = 1, 2 \quad (12.9)$$

where ∇g_i denotes the gradient row vector of $g_i(\mathbf{y})$ at $\mathbf{y}_i^*$. These functions can be written in the compact form

$$\frac{1}{||\nabla g_i||} g_i(\mathbf{y})_{\text{linearized}} = \beta_i - \alpha_i \mathbf{y} \quad \mathbf{i} = 1, 2 \quad (12.10)$$

where $\beta_i = (g_i - \nabla g_i \mathbf{y}_i^*)/||\nabla \mathbf{g_i}||$ is the distance from the origin to the hyperplane $g_i(\mathbf{y})_{\text{linearized}} = 0$ and $\alpha_i = -\nabla \mathbf{g_i}/||\nabla \mathbf{g_i}||$ is the unit normal vector of the hyperplane directed toward the negative values of $g_i(\mathbf{y})$. When $\mathbf{y}_i^*$ is selected on the surface $g_i(\mathbf{y}_i^*) = 0$, we have $\beta_i = \alpha_i \mathbf{y}_i^*$.
3. A first-order approximation of the probability P is obtained by using the linearized surfaces in place of the actual surfaces. Based on well known properties of the standard normal space (Madsen et al [9]),

$$P \approx \Phi(-\beta_1)\Phi(-\beta_2) + \int_0^{\rho_{12}} \phi_{12}(-\beta_1, -\beta_2, \rho)\, d\rho \quad (12.11)$$

where

$$\rho_{12} = \alpha_1 \alpha_2^t$$

$$\phi_{12}(x, y, \rho) = \frac{1}{2\pi\sqrt{1-\rho^2}} \exp\left(-\frac{x^2 + y^2 - 2\rho x y}{2(1-\rho^2)}\right)$$

4. The linearization points should be selected nearest to the point of maximum probability density within F in the standard normal space as the neighborhood of this point contributes most to the probability integral in (12.7). Owing to the radially decaying density of the standard normal space, a single solution for $\mathbf{y}_1^*$ and $\mathbf{y}_2^*$ may be obtained by solving the following constrained optimization program.

$$\text{P 1:} \qquad \text{minimize} \quad ||\mathbf{y}|| \qquad \text{subject to} \quad g_1(\mathbf{y}) \le 0 \text{ and } \mathbf{g_2}(\mathbf{y}) \le 0$$

This solution points lies on the boundary of F, but it may or may not lie on the intersection of the two surfaces $g_1(\mathbf{y}) = \mathbf{0}$ and $g_2(\mathbf{y}) = \mathbf{0}$. If it does, then both the terms $g_i(\mathbf{y}^*)$ are zero. Otherwise, one of these terms is non-zero, say $g_2(\mathbf{y}^*) \neq \mathbf{0}$. In that case, a better linearization point for $g_2(\mathbf{y})$ which lies on the surface $g_2(\mathbf{y}) = \mathbf{0}$ may be chosen. One possible choice is the nearest point on $g_2(\mathbf{y}) = \mathbf{0}$ from $\mathbf{y}_1^*$. Another possibility is to find $\mathbf{y}_2^*$ as the intersection of the vector $\gamma\mathbf{y}_1^*$ with the surface $g_2(\mathbf{y}) = \mathbf{0}$, i.e., solve γ from $g_2(\gamma\mathbf{y}_1^*) = \mathbf{0}$ and set $\mathbf{y}_2^* = \gamma\mathbf{y}_1^*$. Finally, $\mathbf{y}_2^*$ may be obtained as the solution to

$$\text{P 2}: \quad \text{minimize} \quad ||\mathbf{y}|| \qquad \text{subject to} \quad g_1(\mathbf{y}) \leq \mathbf{0} \text{ and } \mathbf{g_2(y)} = \mathbf{0}$$

or

$$\text{P 3}: \quad \text{minimize} \quad ||\mathbf{y}|| \qquad \text{subject to} \quad \beta_1 - \alpha_1\mathbf{y} \leq \mathbf{0} \text{ and } \mathbf{g_2(y)} = \mathbf{0}$$

In P3, we use the linearized version of the first constraint.

To use the above procedure for computing the mean out-crossing rate $\nu(t)$ at time t, we can define the set of random variables as $\mathbf{x} = (\mathbf{u(t)}, \mathbf{u(t+\delta t)}, \mathbf{v})$ and the two limit-state functions as

$$g_1(\mathbf{x}) = -G(\mathbf{u(t)}, \mathbf{v}) \tag{12.12}$$
$$g_2(\mathbf{x}) = G(\mathbf{u(t+\delta t)}, \mathbf{v}) \tag{12.13}$$

However, for small δt, the vectors $\mathbf{u(t)}$ and $\mathbf{u(t+\delta t)}$ are closely correlated. This causes a numerical problem in the required transformation to the standard normal space, as the covariance matrix is close to singularity. An alternative is to define $g_2(\mathbf{x})$ as

$$\begin{aligned} g_2(\mathbf{x}) &= G(\mathbf{u(t)}, \mathbf{v}) + \dot{\mathbf{G}}(\mathbf{u(t)}, \mathbf{v})\delta\mathbf{t} \\ &= G(\mathbf{u(t)}, \mathbf{v}) + \nabla_{\mathbf{u}}\mathbf{G}(\mathbf{u(t)}, \mathbf{v})\dot{\mathbf{u}}\mathbf{(t)}\delta\mathbf{t} \end{aligned} \tag{12.14}$$

Since $G(\mathbf{u(t)}, \mathbf{v})$ is normally an explicit function of $\mathbf{u(t)}$, the gradient $\nabla_{\mathbf{u}}G$ is easy to compute. In this case, the set of random variables is $\mathbf{x} = (\mathbf{u(t)}, \dot{\mathbf{u}}\mathbf{(t)}, \mathbf{v})$. Since $\mathbf{u(t)}$ and $\dot{\mathbf{u}}(t)$ are not closely correlated, this formulation is more expedient from the standpoint of numerical computation.

For a nonlinear system, the joint distribution of $\mathbf{u(t)}$ and $\dot{\mathbf{u}}(t)$ is very difficult or impossible to obtain. In that case, it is better to work with basic variables defining the system properties and the input excitation. Let the random vector $\mathbf{f} = (\mathbf{f_1}, \mathbf{f_2}, ...)^{\mathbf{t}}$ define the input excitation in a discretized form (to be described in the following section) and assume $\mathbf{f}$ and $\mathbf{v}$ are independent. One may symbolically write $\mathbf{u(t)} = \mathbf{u(f, v, t)}$ and $\dot{\mathbf{u}}(t) = \dot{\mathbf{u}}(\mathbf{f}, \mathbf{v}, \mathbf{t})$. These relations are actually in a differential form, as will be developed in the next section. Substituting these relations in (12.12) and (12.14), we have

$$g_1(\mathbf{x}) = -G(\mathbf{f}, \mathbf{v}, \mathbf{t}) \tag{12.15}$$

$$g_2(\mathbf{x}) = G(\mathbf{f}, \mathbf{v}, \mathbf{t}) + \nabla_{\mathbf{u}}\mathbf{G}(\mathbf{f}, \mathbf{v}, \mathbf{t})\dot{\mathbf{u}}(\mathbf{f}, \mathbf{v}, \mathbf{t})\delta\mathbf{t} \tag{12.16}$$

The above relations, of course, are not explicit. However, the basic variables now are $\mathbf{x} = (\mathbf{f}, \mathbf{v})$ whose joint distribution is readily available from the probabilistic specification of the input process and the system parameters. Hence, what we have accomplished is shift the difficult problem of determining the joint distribution of $\mathbf{u(t)}$ and $\dot{\mathbf{u}}(t)$ to that of working with implicit limit-state functions of random variables with known distributions. We have also increased the dimension of the optimization problem, as the size of $\mathbf{x} = (\mathbf{f}, \mathbf{v})$ typically would be much larger than the size of $\mathbf{x} = (\mathbf{u(t)}, \dot{\mathbf{u}}(\mathbf{t}), \mathbf{v})$.

We saw that determination of the linearization points y_1^* and y_2^* requires the solution of an optimization problem. Most efficient algorithms for this purpose require computation of the gradients of the constraint functions (Luenberger [8]; Liu and Der Kiureghian [7]). Hence, we need to determine the gradients of $g_1(\mathbf{x})$ and $g_2(\mathbf{x})$ with respect to $\mathbf{f}$ and $\mathbf{v}$. Assuming that these derivatives exist and using the chain rule of differentiation on (12.15) and (12.16) and the implicit relations between $\mathbf{u(t)}$, $\dot{\mathbf{u}}(t)$, $\mathbf{f}$, and $\mathbf{v}$, we write

$$\nabla_{\mathbf{v}} g_1 = -\left(\nabla_{\mathbf{u}} G \mathbf{J}_{\mathbf{u(t)},\mathbf{v}} + \nabla_{\mathbf{v}}\mathbf{G}|_{\mathbf{u}}\right) \tag{12.17}$$

$$\nabla_{\mathbf{f}} g_1 = -\nabla_{\mathbf{u}} G \mathbf{J}_{\mathbf{u(t)},\mathbf{f}} \tag{12.18}$$

$$\begin{aligned} \nabla_{\mathbf{v}} g_2 = {} & \nabla_{\mathbf{u}} G \mathbf{J}_{\mathbf{u(t)},\mathbf{v}} + \nabla_{\mathbf{v}}\mathbf{G}|_{\mathbf{u}} + \nabla_{\mathbf{u}}(\nabla_{\mathbf{u}}\mathbf{G})\mathbf{J}_{\mathbf{u(t)},\mathbf{v}}\dot{\mathbf{u}}(\mathbf{t})\delta\mathbf{t} + \\ & \nabla_{\mathbf{v}}(\nabla_{\mathbf{u}} G)|_{\mathbf{u}}\dot{\mathbf{u}}(t)\delta t + \nabla_{\mathbf{u}} G \mathbf{J}_{\dot{\mathbf{u}}(\mathbf{t}),\mathbf{v}}\delta\mathbf{t} \end{aligned} \tag{12.19}$$

$$\begin{aligned} \nabla_{\mathbf{f}} g_2 = {} & \nabla_{\mathbf{u}} G \mathbf{J}_{\mathbf{u(t)},\mathbf{f}} + \nabla_{\mathbf{u}}(\nabla_{\mathbf{u}}\mathbf{G})\mathbf{J}_{\mathbf{u(t)},\mathbf{f}}\dot{\mathbf{u}}(\mathbf{t})\delta\mathbf{t} + \\ & \nabla_{\mathbf{u}} G \mathbf{J}_{\dot{\mathbf{u}}(\mathbf{t}),\mathbf{f}}\delta\mathbf{t} \end{aligned} \tag{12.20}$$

In the above, $\mathbf{J}_{\mathbf{p},\mathbf{q}}$ denotes the Jacobian of the transformation $\mathbf{p} = \mathbf{p(q)}$ and the symbol $|_{\mathbf{u(t)}}$ attached to a gradient implies calculation of the gradient with $\mathbf{u(t)}$ fixed.

In the above equations, the gradients of G with respect to $\mathbf{u(t)}$ and with respect to $\mathbf{v}$ with $\mathbf{u(t)}$ fixed are easy to compute, as G is normally an explicit function of $\mathbf{u(t)}$ and $\mathbf{v}$. The main difficulty lies in the computation of the Jacobians $\mathbf{J}_{\mathbf{u(t)},\mathbf{v}}$, $\mathbf{J}_{\mathbf{u(t)},\mathbf{f}}$, $\mathbf{J}_{\dot{\mathbf{u}}(\mathbf{t}),\mathbf{v}}$, and $\mathbf{J}_{\dot{\mathbf{u}}(\mathbf{t}),\mathbf{f}}$. This matter is addressed in the following section.

3 DISCRETE MODEL FOR STOCHASTIC EXCITATION

For implementation of the above method, it is necessary that the input stochastic excitation be described in terms of a vector of random variables. For this purpose and with applications in earthquake engineering in mind, we consider a class of stochastic processes obtained by time-modulation of the response of a filter subjected to a random pulse train. Let $0 = t_0 < t_1 < \cdots < t_n = t$ denote a set of discrete points in time and consider the response of the second-order filter

$$\ddot{u}_f(t) + 2\zeta_f\omega_f\dot{u}_f(t) + \omega_f^2 u_f(t) = \sum_{i=1}^{n} f_i\delta(t - t_i) \tag{12.21}$$

where ω_f and ζ_f are the frequency and damping ratio of the filter, respectively, $\delta(\cdot)$ denotes the Dirac delta function, and f_i are random variables. The right-hand side of the above equation represents a train of random pulses at time instants t_i, $i = 1, 2, ..., n$, with the corresponding random impulse magnitudes f_i, $i = 1, 2, ..., n$. We define f_i as the integral of a white noise process $W(t)$ over the time interval (t_{i-1}, t_i), i.e., $f_i = \int_{t_{i-1}}^{t_i} W(t)dt$. One can easily verify that under this condition f_i are statistically independent, normal random variables with zero mean and variances $2\pi\Phi_0(t_i - t_{i-1})$, where Φ_0 is the intensity of the white noise. Furthermore, for small time increments, the right-hand side represents a discrete idealization of a white noise process. Hence, after a sufficiently long period of time, the response of the filter approaches a stationary process with a filtered white noise power spectral density. In particular, assuming the right-hand side represents the negative of a base acceleration process, the absolute acceleration response of the filter, which is equal to $-(2\zeta_f\omega_f\dot{u}_f(t) + \omega_f^2 u_f(t))$, has a power spectral density approaching the well known Kanai-Tajimi power spectral density (Kanai [5]; Tajimi [14]) often used in describing ground earthquake acceleration. The frequency content of the process can be controlled by proper selection of the filter parameters ζ_f and ω_f, and its intensity can be controlled by selecting Φ_0.

We note that due to the linear nature of the filter, its response in general can be written in the form $\sum_{i=1}^{n} f_i h(t - t_i)$, where $h(t)$ is the unit-impulse response function. For the absolute acceleration response, for example, we have $h(t) = -(2\zeta_f\omega_f\dot{u}(t) + \omega_f^2 u(t))$, where $u(t) = \sin(\omega_d t)\exp(-2\zeta_f\omega_f t)/\omega_d$ with $\omega_d = \omega_f\sqrt{[b]1 - \zeta_f^2}$.

To account for temporal nonstationarity of the process, we multiply to response of the filter by a modulating function $q(t)$. It is also possible to account for spectral nonstationarity. For this purpose we consider the excitation as the sum of the responses of a set of filters with parameters ζ_{f_j} and ω_{f_j}, $j = 1, 2, ..., k$, to the pulse train with each filter response modulated differently. The process then can be written in the form $\sum_{j=1}^{k}\sum_{i=1}^{n} q_j(t) f_i h_j(t - t_i)$, where $h_j(t)$ is the unit-impulse response function for the j-th filter. By proper selection of ζ_{f_j}, ω_{f_j} and $q_j(t)$ almost any form of temporal and spectral nonstationarity can be modeled.

The above model offers many other possibilities. One can consider filters of order other than second, or cascade a series of filters. Furthermore, the filter parameters ζ_{f_j} and ω_{f_j} and the intensity parameter Φ_0 can be considered as random variables to account for uncertainty in the ensemble characteristics of the input excitation. In that case, the set of random variables defining the stochastic excitation are f_i, $i = 1, 2, ..., n$, Φ_0 and ζ_{f_j} and ω_{f_j} for $j = 1, 2, ..., k$. We see that a rather general representation of a nonstationary process is achieved in terms of the set of random variables.

4 GRADIENT OF NONLINEAR RESPONSE

As described earlier, for FORM analysis it is necessary to compute the gradient of the structural response with respect to the random variables. This issue is addressed in this section.

Consider the class of nonlinear dynamic systems with m degrees of freedom whose behavior is governed by the following equation of motion:

$$\mathbf{M}(\mathbf{v},\mathbf{t})\ddot{\mathbf{u}}(\mathbf{t}) + \mathbf{C}(\mathbf{v},\mathbf{t})\dot{\mathbf{u}}(\mathbf{t}) + \mathbf{R}(\dot{\mathbf{u}}(\mathbf{t}),\mathbf{u}(\mathbf{t}),\mathbf{v},\mathbf{t}) = \mathbf{P}\mathbf{F}(\mathbf{t}) \tag{12.22}$$

Here, $\mathbf{M}(\mathbf{v},\mathbf{t})$ denotes the mass matrix, $\mathbf{C}(\mathbf{v},\mathbf{t})$ denotes the viscous damping matrix, $\mathbf{R}(\dot{\mathbf{u}},\mathbf{u},\mathbf{v},\mathbf{t})$ denotes the resisting force vector, $\mathbf{P}$ denotes the load influence matrix, $\mathbf{F}(\mathbf{t})$ denotes a vector of stochastic excitations, and all other variables are as defined earlier. The elements of $\mathbf{F}(\mathbf{t})$ are assumed to be modeled in the manner described in the preceding section. Using the Newmark method of integration (Newmark [10]), we approximate the solution to the above equation for discrete time points $t_0 = 0 < t_1 < \cdots < t_n = t$ (not necessarily the same points as used in defining the input excitation) in a recursive form as follows:

$$\mathbf{u_i} \;=\; ((1-2\beta)\ddot{\mathbf{u}}_{i-1} + 2\beta\ddot{\mathbf{u}}_i)\,(t_i - t_{i-1})^2/2 +$$

$$+\dot{\mathbf{u}}_{i-1}(t_i - t_{i-1}) + \mathbf{u_{i-1}} \tag{12.23}$$

$$\dot{\mathbf{u}}_i = ((1-\alpha)\ddot{\mathbf{u}}_{i-1} + \alpha\ddot{\mathbf{u}}_i)(t_i - t_{i-1}) + \dot{\mathbf{u}}_{i-1} \tag{12.24}$$

where $\mathbf{u_{i-1}}$, $\mathbf{u_i}$, etc., are the responses at time steps t_{i-1} and t_i, respectively, and α and β are the integration parameters.

Using (12.23) and (12.24), for the j-th time step define the nonlinear function vector $\mathbf{Q_j}$

$$\mathbf{Q_j}(\mathbf{v}, \ddot{\mathbf{u}}_\mathbf{j}, \ddot{\mathbf{u}}_\mathbf{j-1}, \dot{\mathbf{u}}_\mathbf{j-1}, \mathbf{u_{j-1}}, \alpha, \beta) = \mathbf{M}(\mathbf{v}, \mathbf{t_j})\ddot{\mathbf{u}}_\mathbf{j} + \mathbf{C}(\mathbf{v}, \mathbf{t_j})\dot{\mathbf{u}}_\mathbf{j} + \tag{12.25}$$

$$\mathbf{R}(\dot{\mathbf{u}}_\mathbf{j}, \mathbf{u_j}, \mathbf{v}, \mathbf{t_j}) - \mathbf{PF}(\mathbf{t_j})$$

Satisfying the equilibrium at the j-th time step requires that

$$\mathbf{Q_j}(\mathbf{v}, \ddot{\mathbf{u}}_\mathbf{j}, \ddot{\mathbf{u}}_\mathbf{j-1}, \dot{\mathbf{u}}_\mathbf{j-1}, \mathbf{u_{j-1}}, \alpha, \beta) = \mathbf{0} \tag{12.26}$$

Let x be a random variable which is either an element of the system parameters $\mathbf{v}$ or one of the random variables defining the input excitation process (f_i, ζ_{f_j}, ω_{f_j} or Φ_0). Assuming $\mathbf{Q_j}$ is differentiable, the gradients of the nonlinear response are:

$$\frac{\partial \ddot{\mathbf{u}}_j}{\partial x} = -\left(\frac{\partial \mathbf{Q_j}}{\partial \ddot{\mathbf{u}}_\mathbf{j}}\right)^{-1} \times \tag{12.27}$$

$$\left(\frac{\partial \mathbf{Q_j}}{\partial \mathbf{x}} + \frac{\partial \mathbf{Q_j}}{\partial \mathbf{u_{j-1}}}\frac{\partial \mathbf{u_{j-1}}}{\partial \mathbf{x}} + \frac{\partial \mathbf{Q_j}}{\partial \dot{\mathbf{u}}_\mathbf{j-1}}\frac{\partial \dot{\mathbf{u}}_{j-1}}{\partial x} + \frac{\partial \mathbf{Q_j}}{\partial \ddot{\mathbf{u}}_\mathbf{j-1}}\frac{\partial \ddot{\mathbf{u}}_{j-1}}{\partial x}\right)$$

$$\frac{\partial \dot{\mathbf{u}}_j}{\partial x} = \left((1-\alpha)\frac{\partial \ddot{\mathbf{u}}_{j-1}}{\partial x} + \alpha\frac{\partial \ddot{\mathbf{u}}_j}{\partial x}\right)(t_j - t_{j-1}) + \frac{\partial \dot{\mathbf{u}}_{j-1}}{\partial x} \tag{12.28}$$

$$\frac{\partial \mathbf{u_j}}{\partial \mathbf{x}} = \left((1-2\beta)\frac{\partial \ddot{\mathbf{u}}_{j-1}}{\partial x} + 2\beta\frac{\partial \ddot{\mathbf{u}}_j}{\partial x}\right)(t_j - t_{j-1})^2/2 +$$

$$\frac{\partial \dot{\mathbf{u}}_{j-1}}{\partial x}(t_j - t_{j-1}) + \frac{\partial \mathbf{u_{j-1}}}{\partial \mathbf{x}} \tag{12.29}$$

When $x = f_i$, clearly $\partial \mathbf{u_j}/\partial \mathbf{f_i}$, $\partial \dot{\mathbf{u}}_j/\partial f_i$, and $\partial \ddot{\mathbf{u}}_j/\partial f_i$ are zero for $i > j$. From the above recursive relations, the Jacobians $\mathbf{J_{u_j,x}}$, $\mathbf{J_{\dot{u}_j,x}}$, and $\mathbf{J_{\ddot{u}_j,x}}$ are obtained in terms of the previous Jacobians $\mathbf{J_{u_{j-1},x}}$, $\mathbf{J_{\dot{u}_{j-1},x}}$, and $\mathbf{J_{\ddot{u}_{j-1},x}}$, and the current values of $\mathbf{u_j}$, $\dot{\mathbf{u}}_j$, and $\ddot{\mathbf{u}}_j$.

The conditions under which $\mathbf{Q_j}$ is differentiable depends on the characteristics of the nonlinear system and the response of interest. For the first two examples

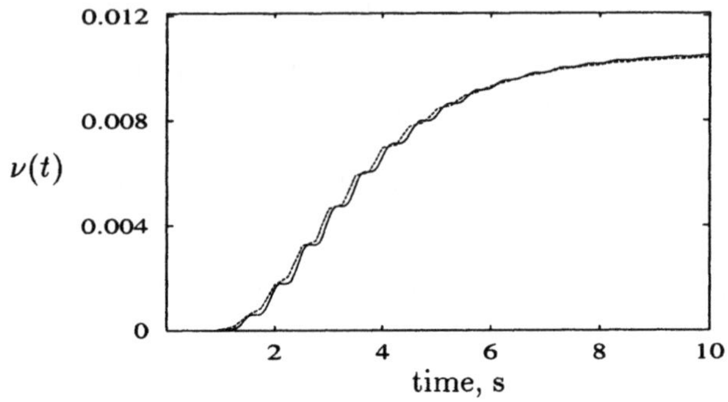

Figure 1 Mean out-crossing rate of linear oscillator subjected to white noise.

considered in the following section, $\mathbf{Q_j}$ is differentiable. For the third example, the hysteretic oscillator, strictly speaking $\mathbf{Q_j}$ is not differentiable. However, with a small modification in the equations of motion, it is made to be differentiable.

5 EXAMPLES

To demonstrate the proposed method, three example systems are considered: a linear oscillator subjected to white noise, a duffing oscillator subjected to white noise and filtered white noise, and a hysteretic oscillator subjected to filtered white noise. Where possible, the results by the proposed method are compared with exact or simulation results.

5.1 Linear Oscillator

The equation of motion is given by

$$\ddot{u}(t) + 2\omega\zeta\dot{u}(t) + \omega^2 u(t) = F(t) \tag{12.30}$$

where ω denotes the natural frequency and ζ denotes the damping ratio of the oscillator.

For $\omega = 6$ rad/s, $\zeta = 0.05$, and $F(t)$ a suddenly applied white noise of intensity $\Phi_0 = 1$, Fig. 1 shows plots of the mean out-crossing rate for the limit-state function $G = 3\sigma - u(t)$. The solid line is based on the exact result (Nigam [11]), and the dashed curve is computed by the proposed method. For the latter, the white noise input is represented as a pulse train with $t_i - t_{i-1} = 0.02$s in the manner described in Section 3. The agreement between the two results is remarkable. This is not surprising, since for this case the limit-state function in the space of the random variables f_i is linear and, hence, no approximation due to linearization of the limit-state surface is introduced. Small discrepancies between the two sets of results are due to discretization of the input process.

5.2 Duffing Oscillator

The equation of motion is given by

$$\ddot{u}(t) + 2\omega\zeta\dot{u}(t) + \omega^2(1 + \gamma u^2(t))u(t) = F(t) \tag{12.31}$$

where ω, ζ and γ are the system parameters, with the latter representing a measure of nonlinearity of the oscillator.

We first consider the duffing oscillator subjected to white noise input, which is idealized as a train of random pulses in the manner described above. For the system parameters, $\omega = 20$ rad/s, $\zeta = 0.1$, and the range $\gamma = 0.2$-0.8 are considered. The oscillator is assumed to be initially at rest. The linearization point for implementing the FORM approximation is obtained by employing the P3 optimization formulation and $\delta t = 0.0001$s is used for computing equation (12.6). $2\beta = \alpha = 0.5$ is used for the numerical integration scheme.

For $\gamma = 0$, the system is linear and the stationary mean-square response is $\sigma^2 = \pi\Phi_0/2\zeta\omega^3$. We examine the out-crossings of the oscillator for the limit-state function $G = 3\sigma - u(t)$ (equivalent to up-crossings of the 3σ level of the linear oscillator) for different levels of nonlinearity. Fig. 2a shows plots of the computed mean out-crossing rate as a function of time for γ ranging from 0.2 to 0.8. The mean out-crossing rate sharply decreases with increasing nonlinearity due to the increasing stiffness of the oscillator. We see that beyond $t = 1.5$s the rates are essentially constant, indicating near stationarity of the response beyond this point. These "stationary" values are compared with exact values (Lin [6]) in Table 1. Also compared are stationary mean out-crossing rates computed using the equivalent linearization method (Caughey [3], Atalik et al [1]). It is evident that the accuracy of the proposed method is far superior to that of the equivalent linearization method.

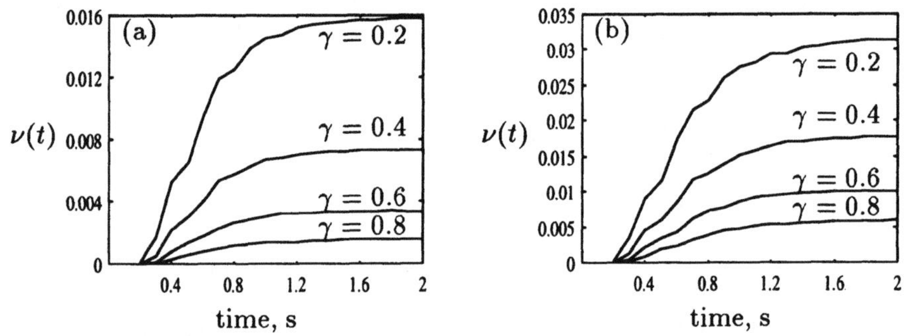

Figure 2 Mean out-crossing rate of duffing oscillator with (a) deterministic and (b) uncertain properties subjected to white noise.

γ	exact value	proposed method	equivalent linearization
0.2	1.64×10^{-2}	1.58×10^{-2}	2.30×10^{-2}
0.4	7.55×10^{-3}	7.34×10^{-3}	1.60×10^{-2}
0.6	3.47×10^{-3}	3.36×10^{-3}	1.15×10^{-2}
0.8	1.59×10^{-3}	1.54×10^{-3}	8.53×10^{-3}

Table 1 Comparison of computed stationary mean out-crossing rates.

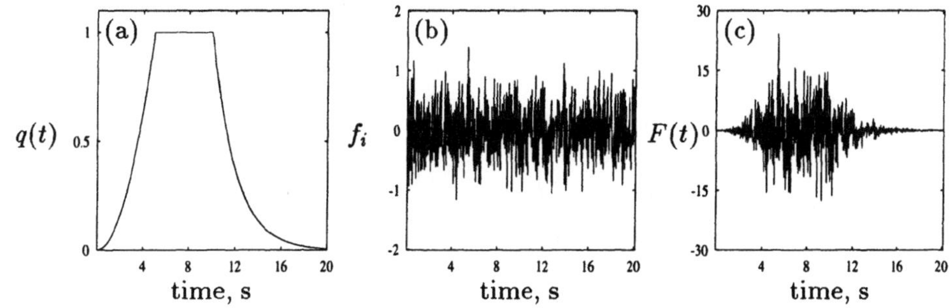

Figure 3 (a) Time-modulating function $q(t)$, (b) random sample of pulse train f_i, (c) random sample of the excitation $F(t)$

To examine the effect of uncertainty in system properties, we assume ω and ζ are lognormal random variables with means equal to the values specified earlier and coefficients of variation equal to 0.05 and 0.20, respectively. The computed mean out-crossing rates based on the proposed method are shown in Fig. 2b. Corresponding exact or equivalent linearized results, of course, do not exist. Comparison with the results in Fig. 2a shows that the mean out-crossing rate sharply increases in account of the system uncertainty. This is due to the likelihood of having a lower stiffness when ω is random.

Next, the excitation $F(t)$ is assumed to be a modulated, filtered white noise process of intensity $\Phi_0 = 1$ and modulating function

$$\begin{aligned} q(t) &= t^2/25 && t \leq 5 \\ &= 1 && 5 < t \leq 10 \\ &= \exp(-(t-10)/2) && t > 10 \end{aligned}$$

Filter parameters are assumed to be $\omega_f = 5\pi$ rad/s and $\zeta_f = 0.6$.

The process is idealized by a pulse train in the manner described in Section 3 using constant time increments $t_i - t_{i-1} = 0.025$s. Fig. 3 shows a plot of the modulating function and random samples of the pulse train and the absolute acceleration response of the filter. The latter is a realization of the stochastic input into the duffing oscillator. The process is seen to be a realistic representation of ground acceleration caused by an earthquake.

For the present example the oscillator frequency ω and damping ratio ζ are assumed to be lognormal random variables with means 6 rad/s and 0.1 and

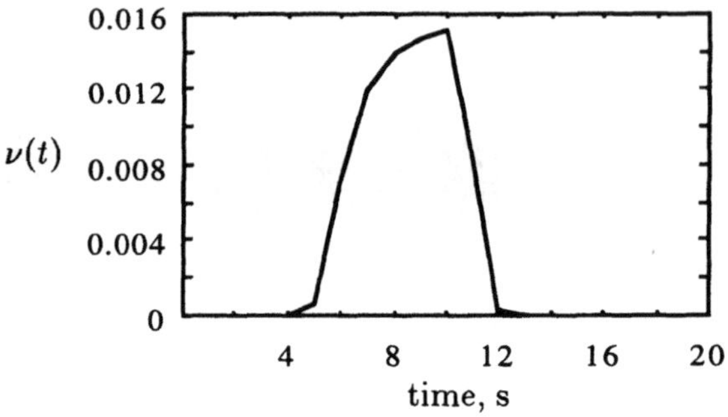

Figure 4 Mean out-crossing rate of duffing oscillator with uncertain properties and subjected to modulated filtered white noise.

coefficients of variation 0.05 and 0.20, respectively. The nonlinearity parameter is assumed to be $\gamma = 0.2/\sigma^2$, where σ^2 denotes the mean-square stationary response of the linear oscillator ($\gamma = 0$) with mean properties ($\omega = 6$ rad/s and $\zeta = 0.1$). We examine the out-crossings of the oscillator for the limit-state function $G = 2.5\sigma - u(t)$. The mean out-crossing rate of the uncertain oscillator subjected to the non-white, nonstationary excitation is shown in Fig. 4. The mean out-crossing rate is seen to increase rapidly after $t = 5$s and, after reaching a near flat plateau at $t = 10$s, it rapidly decreases due to the decay in the intensity of the excitation. The area underneath the mean out-crossing curve, $p = 0.0567$, represents an upper bound to the probability of failure, i.e., the probability that the nonlinear response will exceed the level 2.5σ. This result compares favorably with a Monte Carlo simulation result of $p_f = 0.0479$, which is obtained with a coefficient of variation of 0.02 by 52,000 simulations.

An important advantage of the proposed method is the richness of the information gained from the solution. Shown in Fig. 5 are plots of oscillator frequency ω and damping ratio ζ at the linearization point (the so called "design point") of the FORM analysis as functions of time. These can be seen as the most likely values of these oscillator parameters that may give rise to the out-crossing event at each time instant. We see that the "design point" values for both parameters are below their mean values, particularly when the input excitation is weak. This is because the out-crossing event is more likely to occur when the oscillator is flexible (small ω value) and lightly damped (small ζ value).

Fig. 5 shows a plot of the "design point" excitation for the out-crossing event at $t = 10$s. Among all possible realizations of the input excitation, this is the one

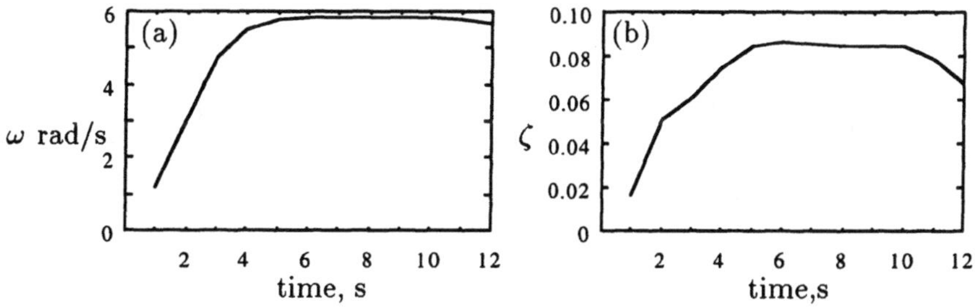

Figure 5 Design points values of (a) frequency ω and (b) damping ratio ζ.

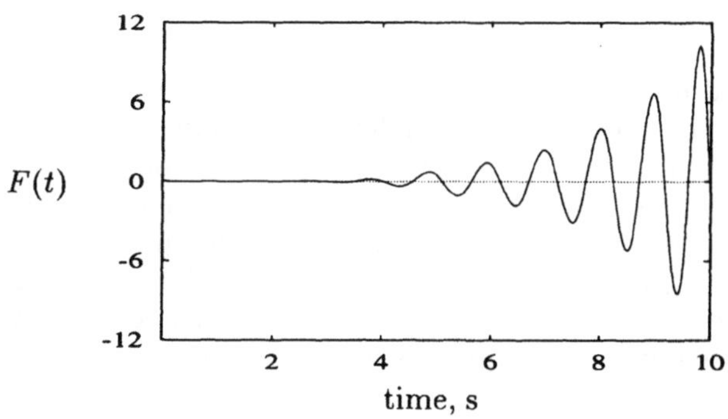

Figure 6 Design point excitation for $t = 10$.

that has the highest likelihood if the out-crossing event is to occur at $t = 10$s. It is useful to note that the "design point" values of the input excitation random variables f_i for $t < 4$s are nearly zero. This indicates that the out-crossing event at $t = 10$s is essentially independent of the input excitation during $0 < t < 4$s.

5.3 Hysteretic Oscillator

Consider the equations of motions of an oscillator having an hysteretic restoring force in accordance to the well known Bouc-Wen model (Bouc [2]; Wen [15])

$$\ddot{u}(t) + 2\zeta\omega\dot{u}(t) + \omega^2(\alpha u(t) + (1-\alpha)z(t)) = F(t) \qquad (12.32)$$

$$\dot{z}(t) = A\dot{u}(t) - \gamma|\dot{u}(t)||z(t)|^{n-1}z(t) - \beta\dot{u}(t)|z(t)|^n \qquad (12.33)$$

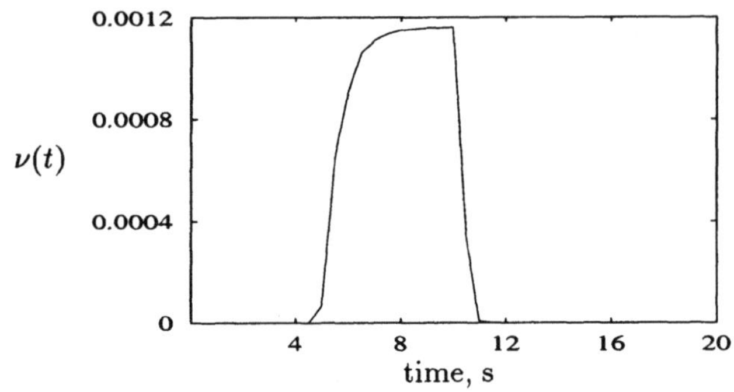

Figure 7 Mean out-crossing rate of hysteretic oscillator subjected to modulated filtered white noise.

where ω, ζ, and α are system parameters with α controlling the measure of nonlinearity of the oscillator ($\alpha = 1$ for linear system), $z(t)$ is an auxiliary variable, and γ, n, β and A are parameters that control the shape of the hysteresis loop. The response of this oscillator is not differentiable due to the presence of absolute value terms in the second equation. We resolve this problem by making a small modification in the second equation. The details are beyond the scope of this paper and will not be presented here. The excitation $F(t)$ is the filtered white noise process described above.

For $\omega = 6$ rad/s, $\zeta = 0.05$, $\alpha = 0.2$, $\gamma = 1/(2\sigma)$, $n = 1$, $\beta = 1/(2\sigma)$ and $A = 1$, Fig. 7 shows a plot of the mean out-crossing rate for the limit-state function $G = 3\sigma - u(t)$, where σ denotes the root-mean-square response for the linear ($\alpha = 1$) oscillator. The area underneath the curve equals 0.0055. This value closely bounds the failure probability $p_f = 0.00511$ obtained by 78,500 simulations with a coefficient of variation of 0.0498.

To highlight the nonlinearity of the response, Fig. 8 shows the probability density function of the oscillator response at $t = 10$s for $\alpha = 1$, 0.6 and 0.2, the first being the case for a linear oscillator. This is obtained as the negative of the sensitivity of the failure probability for the limit-state function $G = u_0 - u(10)$ with respect to u_0. It is seen that, for the selected values of ω and ζ, the probability mass of the nonlinear response moves towards the origin. This is due to the fact that the hysteretic oscillator has an effective natural frequency that is smaller than $\omega = 6$ rad/s, which is at more distance from the dominant range of input frequencies ($\omega_f = 5\pi$ rad/s) than for the linear oscillator. The three distributions are also plotted on a normal probability paper in Fig. 9. It

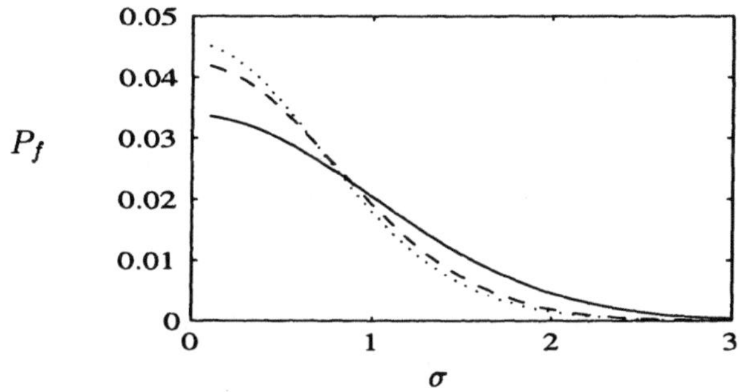

Figure 8 Probability density function of the hysteretic response for $\alpha = 1$ (solid line), $\alpha = 0.6$ (dashed line), and $\alpha = 0.2$ (dotted line).

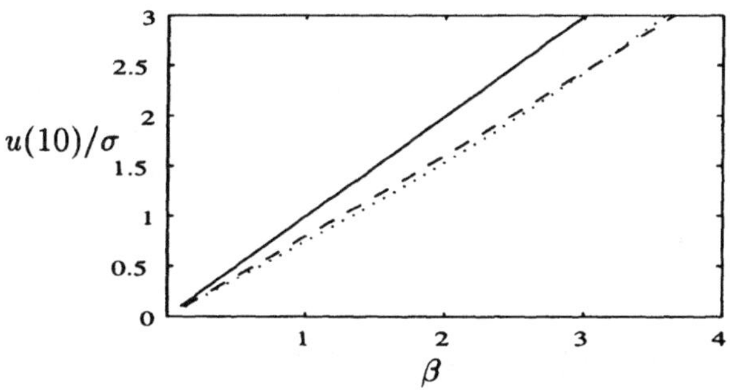

Figure 9 Normal probability paper (solid line for $\alpha = 1$, dashed line for $\alpha = 0.6$, and dotted line for $\alpha = 0.2$).

can be seen that, whereas for the linear oscillator (solid line) the distribution plots as a straight line and therefore is Gaussian, for the hysteretic oscillators the distributions do not plot as straight lines (note that the lines cross at two points) and therefore are not Gaussian. This is a good indication of the nonlinearity of the response.

6 CONCLUSION

A new method for computing the mean out-crossing rate of nonlinear uncertain systems subjected to stochastic excitation is developed. The method is general and can be applied to a broad class of nonlinear systems with multiple degrees of freedom. Any uncertainty in the characteristics of the system or the input excitation can be easily incorporated in the analysis. The mean out-crossing rate can be used to obtain an upper bound on the probability of failure.

A class of nonstationary, non-white stochastics processes characterized by a vector of random variables is modeled to allow implementation of the method, which is based on the first-order reliability method (FORM). It is found that this representation is sufficiently rich to allow arbitrary forms of temporal or spectral nonstationarity in the excitation.

For the considered linear, duffing, and hysteretic oscillators subjected to white or filtered white noise input, the mean out-crossing rates computed by the proposed method are in excellent agreement with exact or simulation results. Besides the mean out-crossing rate, the method provides information on the most likely characteristics of the system and the excitation in the event of out-crossing. Such information provides valuable insight into understanding the behavior of a complex system.

7 ACKNOWLEDGEMENT

This research was sponsored by the CUREe-Kajima grant M1975. This support is gratefully acknowledged.

REFERENCES

[1] T. S. Atalik and S. Utku. Stochastic linearization of multi-degree-of-freedom non-linear systems. *Earthquake Engng. Struct. Dyns.*, 4:411–420, 1976.

[2] R. Bouc. Forced vibration of mechanical systems with hysteresis. In *Proceedings of the Fourth Conference on Nonlinear Oscillation*. Prague, Czechoslovakia, 1967.

[3] T. K. Caughey. Equivalent linearization techniques. *J. Acoust. Soc. Amer*, 35:1706–1711, 1963.

[4] Ø. Hagen and L. Tvedt. Vector process out-crossing as parallel system sensitivity measure. *Journal of Engineering Mechanics*, 117:2201–2220, 1991.

[5] K. Kanai. Semi-empirical formula for the seismic characteristics of the ground. *Univ. Tokyo Bull. Earthquake Res. Inst.*, 35:309–325, 1957.

[6] Y. K. Lin. *Probabilistic theory of structural dynamics*. McGraw-Hill Inc., New York, 1967.

[7] P. L. Liu and A. Der Kiureghian. Optimization algorithms for structural reliability. *Structural Safety*, 9:161–177, 1991.

[8] D. G. Luenberger. *Linear and nonlinear programming*. Addison-Wesley, Reading, Massachusetts, 2nd edition, 1984.

[9] H. O. Madsen, S. Krenk, and N. C. Lind. *Methods of structural safety*. Prentice-Hall Inc., Englewood Cliffs, New Jersey, 1986.

[10] N. M. Newmark. A method of computation for structural dynamics. *Journal of Engineering Mechanics*, 85:67–94, 1959.

[11] N. C. Nigam. *Introduction to random vibrations*. The MIT Press, Cambridge, Massachusetts, 1983.

[12] S. O. Rice. Mathematical analysis of random noise. *The Bell System Technical Journal*, 23:282–332, 1944.

[13] S. O. Rice. Mathematical analysis of random noise. *The Bell System Technical Journal*, 24:46–156, 1945.

[14] H. Tajimi. A statistical method of determining the maximum response of a building structure during a earthquake. *Proc. 2d World Conf. Earthquake Eng. Tokyo and Kyoto*, 2:781–798, 1960.

[15] Y. K. Wen. Approximate method for nonlinear random vibration. *Journal of Engineering Mechanics*, 102:249–263, 1976.

13

REVIEW PROBLEM OWNER PERSPECTIVE

Arie Kraak

Road and Hydraulic Engineering Division,
Rijkswaterstaat, Delft, The Netherlands

This contribution is written from the perspective of a problem owner, and is therefore not focussed on the mathematics but is merely meant, by means of two questions, to trigger the authors to give insight in the applicability of their proposed method.

In the paper under discussion a mathematical solution is proposed to calculate the probability of exceedance of a certain response (in terms of displacements for instance) of a multi degree of freedom system when exposed to a stochastic load. My main interest lies in the applicability for flood defence problems, however a side step towards the offshore industry might be interesting. Oil companies such as Shell and Exxon are exploring oil in the North-Sea. Often they use for exploration purposes so called Jack-up platforms. These platforms are used in varying but relatively shallow water (up to 100 m depth) and on different spots in relative short time. For this reason these Jack-ups are easy to transport (they float like a ship) and consist of a platform which can be pulled along their legs which are founded on the sea bed. In this way the platform can be adjusted to the local waterdepth. However the platform shows dynamic responses to wave loading, due to its overall weak stiffness. One of the key questions for an oil company is the reliability of the structure during life time. Apart from fatigue due to frequent loading, this will depend on the probability that a certain excitating force will cause untolerable forces in the structure. Large numerical models are available (sometimes only at the oil company itself), which calculate the dynamic response of the system to a certain stochastic waveforce, both in time domain and frequency-domain. The platform is schematised in hundreds of nodes and branches with determinstic properties. The system is a multi-degree of freedom system with dynamical behaviour and with large uncertainties in:

R. Cooke et al. (eds.), Engineering Probabilistic Design and Maintenance for Flood Protection, 161–162.

- structural damping;
- structural stiffness;
- foundation stiffness;
- drag- and inertiacoefficients in the Morison-equation;
- structure-fluid interaction, such as added mass and damping.

How can the proposed method presented in the paper be incorporated in these large numerical but so far deterministic models?

Now, I want to focus on design problems of flood defences. Several mechanisms may cause failure of a dyke, dune or othe type of flood defence. Some mechanisms are strongly correlated to the (extreme) waterlevels, such as overtopping of a dyke, others are less correlated, such as geotechnical failure. I will focus on the first type. In dyke design the average amount of water which is allowed to flow over the dyke during overtopping under extreme conditions is limited. Depending of the quality of the inner-slope only 0.1, 1.0 or 10.0 l/m/s is allowed to flow over a dyke-crest. Until now the duration of the overtopping event is not taken into account in dyke design, or in other words the dyke is considered to be safe during infinite time under design-conditions. However in future, the duration of overtopping events (and other failure mechansms) will be taken into account as well in order to optimise design criteria. My second question is:

Can the presented model be used for this type of problem in which the duration of exceedance of a certain level of excitation' seems of importance, more than the event itself?

14

REVIEW ENGINEERING PERSPECTIVE

Ton Vrouwenvelder

Department of Civil Engineering,
Delft University of Technology, Delft, The Netherlands

It is a difficult task to express "the" engineer's view to the above mentioned paper. The problem is: which engineer is supposed to give a view? I know many civil engineers who would qualify papers of this type simply as "not understandable," as "scientific humbug," or as "not relevant for the every day design work." I think that only a small group of engineers would recognize the relevance of this type of studies and understand the link between the mathematical work on the one side and a good and justified practical design on the other. Even of that group only a small number would be able to understand the paper in detail. However, being in that subset, it is a pleasure to read the paper. It has been well written and it summarises in a condensed but clear way a great part of the necessary theoretical reliability background.

Let me now come to a more detailed comment. I do not agree with the claim in the paper that there is not a good and simple linearisation method. Consider the nonlinear spring force of equation (31):

$$F_{spring} = \omega^2(1 + \gamma u^2)u.$$

A generally reliable method to define an equivalent linear spring with stiffness $k\omega^2$ is to tune the amount of stored elastic energy:

$$\int F_{spring}\, du = \frac{1}{2}\omega^2 k u^2.$$

Now:

$$\int F_{spring}\, du = \int \omega^2(1 + \gamma u^2)u\, du = \omega^2 \left(\frac{1}{2}u^2 + \frac{1}{3}\gamma u^3\right).$$

So finally:

$$k = 1 + \frac{2}{3}\gamma u.$$

R. Cooke et al. (eds.), Engineering Probabilistic Design and Maintenance for Flood Protection, 163–165.

The value of u we are interested in is the $u_L = 3\sigma_u$ level of the linear case, where σ_u is given by:

$$\sigma_u = \sqrt{\frac{\pi\phi_0}{2\zeta\omega^3}} = 0.44.$$

So, with $\gamma = 0.8$ (most nonlinear case) and $u_L = 3\sigma_u = 1.33$ we have for k:

$$k = 1 + \frac{2}{3}\gamma u = 1 + \frac{2}{3}0.8 * 1.33 = 1.71.$$

For the linear case ($\gamma = 0$, $\omega = 20$ rad/s, $\beta = u_L/\sigma_u = 3$) we simply have the well known expression for the outcrossing rate:

$$\nu = \frac{\omega}{2\pi}\exp(-\beta^2/2) = \frac{20}{2\pi}\exp(-9/2) = 0.035.$$

This value is not given in the paper, but it seems correct if we extrapolate Figure 2a or the "exact value column" in Table 1.

In the nonlinear case for $\gamma = 0.8$ we have the following linearized differential equation:

$$\ddot{u} + c\dot{u} + k\omega^2 u = F(t).$$

The corresponding σ_u-value can now best be written as:

$$\sigma_u = \sqrt{\frac{\pi\phi_0}{ck\omega^2}}.$$

In the nonlinear case k has changed but not c and ω. So now we have:

$$c = 2\zeta\omega = 2 * 0.1 * 20 = 4,$$

$$\sigma_u = \sqrt{\pi * 100/4 * 1.71 * 20^2} = 0.34,$$

$$\beta = u_L/\sigma_u = 1.33/0.34 = 3.9.$$

The expression for the outcrossing rate ν should be modified as follows:

$$\nu = \frac{\omega\sqrt{k}}{2\pi}\exp(-\beta^2/2) = \frac{20 * \sqrt{1.71}}{2\pi}\exp(-2 * 3.9^2) = 2.07 \times 10^{-3}.$$

This numerical result is not far from the exact value of 1.59^{-3} in table 1.

Given the above simplified calculation model it is also possible to make a fair estimate for the reliability of the oscillator with random properties. The advantage of such a procedure is that it gives the possibility to check the more rigorous calculations given in the paper. Especially for an "engineer" it is important to keep the feeling that the theory and the calculations are still under control. Advanced mathematical theories and corresponding operational computer programs are nice and indispensable for large problems. However, an engineer should also have the opportunity to understand and verify these problems in rough lines.

15

REVIEW MATHEMATICAL PERSPECTIVE

Arne Huseby

Department of Mathematics, University of Oslo, Oslo, Norway

1 INTRODUCTION

We start out by reviewing the basic terminology and notation used in the paper. The authors consider what is called a response process, denoted by $\mathbf{u(t)}$ where t denotes the time. It is assumed a finite time horizon, i.e., $0 \leq t \leq T$. A vector $\mathbf{v}$ denotes a set of system variables affecting the process in a certain way. Apparently $\mathbf{v}$ is considered to be constant over time. Depending on the system variables and the response process, the system may be either in a safe or an unsafe state. These two states are defined by introducing a limit-state function, $G(\mathbf{u(t)}, \mathbf{v})$. The process is said to be in an unsafe state if $G < 0$. When the process enters the unsafe area, this is called an out-crossing. The problem considered in the paper is to calculate the probability of at least one out-crossing in a certain time interval. That is, the objective is to calculate:

$$P_f(T) = P(\{\min_{0 \leq t \leq T} G(\mathbf{u(t)}, \mathbf{v})\} \leq \mathbf{0}). \tag{15.1}$$

In order to relate this problem to an application of relevance to flood risks, we could e.g., interpret $\mathbf{u(t)}$ as the water levels at certain locations at time t. The system variables $\mathbf{v}$ may be parameters in the socalled Chezy equations describing the water flow at these locations. Finally, we may think of the limit-state function G as defining the design levels of a flood protection system.

R. Cooke et al. (eds.), Engineering Probabilistic Design and Maintenance for Flood Protection, 167–172.

2 RISK MEASURES

It is natural to look at (15.1) as a risk measure. We notice that the main concern behind this measure is whether or not the process enters the unsafe state during the period of interest. Other aspects such as how long time the process spends in the unsafe state, how far into the unsafe area the process moves, etc., are not taken into account. Hence, it is probably most useful in situations where the negative effects of entering the unsafe state are extremely severe, and something one would want to avoid almost at any cost.

In less extreme settings one may want to consider other measures capturing different aspects. At this stage we would like to present some other alternatives.

One possible option is to consider the following measure:

$$E[\int_0^T \mathrm{I}(G(\mathbf{u(t)}, \mathbf{v}) \leq \mathbf{0})\mathbf{dt}]. \tag{15.2}$$

That is, the expected time the process spend in the unsafe state. This type of measure is useful when the negative effects of being in the unsafe state accumulates over time. If you already lost your house in a flood, it probably does not matter that much if the flood continues yet another week, assuming of course that you have a nice, dry place to stay during this week. However, if you are running a factory that is shut down because of a flood, then the longer the flood lasts, the more you lose in production. In such cases the measure (15.2) is more useful.

As stated, the main purpose of the limit-state function, G is to define the boundary between the safe and unsafe states. Apart from that G does not necessarily have any physical meaning. Sometimes, however, G does have a specific practical interpretation. In such situations it may be of interest to know more than just whether or not G is negative. The actual value of G may be important too. A possible risk measure may then be:

$$E[\int_0^T \max\{-G(\mathbf{u(t)}, \mathbf{v}), \mathbf{0}\}\mathbf{dt}]. \tag{15.3}$$

If the G function measures the amount of damage in the unsafe area, this measure can be interpreted as the expected cumulative damage in the period of interest.

In order to explore possible effects of an out-crossing, it is often convenient to condition on the occurrence of an out-crossing event. Assuming e.g., that the process has an out-crossing at time t_0, we may introduce:

$$X = \inf\{t > t_0 : G(\mathbf{u}(\mathbf{t}), \mathbf{v}) > \mathbf{0}\}. \tag{15.4}$$

Thus, X is the duration of the period where the process stays in the unsafe state. Then it may be of interest to calculate e.g., the distribution of X, $E(X)$, or $E[c(X)]$ for some suitable cost function $c(x)$.

Which risk measure one should use, will obviously depend on the situation. The proposed measure may be relevant in certain settings. Other times one would choose alternative measures. Anyway it would be interesting to see if the approximations used in the paper could be applied in order to calculate some of the other measures as well.

3 A COUNTING PROCESS APPROACH

We now turn to a more detailed discussion about the probability model used in the paper. A natural way to attack the problem is to use a counting process approach. To do so we introduce:

$$N(t) = \text{The number of out-crossings in } (0, t], \tag{15.5}$$

and make the following assumptions:

$$P(N(t+h) = n+1 | N(t) = n) = \nu_n(t)h + o(h) \tag{15.6}$$

$$P(N(t+h) = n | N(t) = n) = (1 - \nu_n(t)h) + o(h) \tag{15.7}$$

$$P(N(t+h) > n+1 | N(t) = n) = o(h) \tag{15.8}$$

Notice that in this model we allow the out-crossing rate at time t to depend on the number of out-crossings in $(0, t]$. From an applied point of view, this flexibility is often important. In particular, this is typically true when dealing with floods. It is well-known that a flood may change the flow characteristics of a river. Because of this, it is common practice to update the parameters in the Chezy equation frequently.

Solving the differential equations derived from the assumptions (15.6), (15.7) and (15.8), we get:

$$P_f(T) = P(N(T) \geq 1) = 1 - P(N(T) = 0) = 1 - \exp(-\int_0^T \nu_0(t)dt) \tag{15.9}$$

If the integral in the exponent is small, it follows by Taylor approximation that:

$$P_f(T) \approx 1 - (1 - \int_0^T \nu_0(t)dt) = \int_0^T \nu_0(t)dt. \qquad (15.10)$$

In fact (15.10) is an upper bound on $P_f(T)$. The above serves as a motivation for calculating the out-crossing rate $\nu_0(t)$.

In order to use the methods proposed in the paper, we need to make the assumption that the out-crossing rate at time t does not depend on the number of out-crossings in $(0, t]$. This implies that we will not be able to handle system changes due to out-crossings. As already mentioned, this assumption is hardly realistic in the case of floods. It should be noted, however, that at least in low risk situations this issue is less important since the probability of having more than one out-crossing in an interval of time is neglectable.

Under the above assumption we may change the notation and simply denote the out-crossing rate by $\nu(t)$ instead of $\nu_0(t)$. Moreover, $\nu(t)$ can be calculated using the following expression:

$$\nu(t) = \lim_{\Delta t \to 0} \frac{P(g_1 < 0 \cap g_2 < 0)}{\Delta t}, \qquad (15.11)$$

where:

$$g_1 = -G(\mathbf{u(t)}, \mathbf{v}) \qquad (15.12)$$

and:

$$g_2 = G(\mathbf{u(t + \Delta t)}, \mathbf{v}). \qquad (15.13)$$

Expressing $\nu(t)$ as in (15.11) is done to facilitate the use of FORM approximation.

4 FORM APPROXIMATION

For our purpose we may describe FORM approximation as a method for solving the following very general problem: Let $\mathbf{x}$ be a random vector and g_1 and g_2 be two limit-state functions. No further assumptions are made concerning the distribution of $\mathbf{x}$ or the two limit-state functions. The fact that so few assumptions are made, makes the proposed method very general. The quality of the approximation, however, may be unacceptable in certain situations.

As indicated above there are two issues we need to handle in the FORM approximation. The first issue is how to deal with general distributions. The answer to this is to use the following representation: Let $\mathbf{y}$ be a vector of independent standard normally distributed random variables. We then try to represent $\mathbf{x}$ as follows:

$$\mathbf{x} = \mathbf{x}(\mathbf{y}) \tag{15.14}$$

In principle it is always possible to do this. In the univariate case we may e.g., represent an arbitrary random variable x as:

$$x = F^1(\Phi(y)), \tag{15.15}$$

where F is the cdf. of x and Φ is the cdf. of a standard normal distribution. Note, however, that representing a vector $\mathbf{x}$ as $\mathbf{x}(\mathbf{y})$ does not imply that the inverse transformation exists. If e.g., x is discrete, then it is not possible to transform x into a normally distributed y. Still in the applications considered in the paper, this probably not a problem.

The second issue is how to handle general limit-state functions. The solution is to approximate the regions:

$$\{\mathbf{y} : \mathbf{g_i}(\mathbf{x}(\mathbf{y})) < \mathbf{0}\},\ \mathbf{i} = \mathbf{1}, \mathbf{2}, \tag{15.16}$$

by linearized regions:

$$\{\mathbf{y} : \ell_{\mathbf{i}}(\mathbf{x}(\mathbf{y})) < \mathbf{0}\},\ \mathbf{i} = \mathbf{1}, \mathbf{2}, \tag{15.17}$$

where ℓ_1 and ℓ_2 are linear functions, i.e., functions of the following form:

$$\ell_i(\mathbf{y}) = \beta_{\mathbf{i}} - \boldsymbol{\alpha}_{\mathbf{i}}\mathbf{y} \tag{15.18}$$

Notice that the linearization is done in $\mathbf{y}$-space, not in $\mathbf{x}$-space.

To obtain good approximations it is very important to choose the right linearization points. The paper suggests some methods for doing this. To make these work, it is implicitely assumed that the origin does not belong to any of the regions (15.16). Moreover, one need to assume that no region is a subset of the other. The main idea behind the methods is that the linear approximations should be good in areas with high probability, i.e., areas close to the origin. Apart from this idea, the proposed methods appears to be somewhat ad. hoc. It is hard to tell how well they would work in such a general setup.

At least in a case where the regions (15.16), and hence their intersection as well, are convex, it is possible to do better. In this particular case any linear approximation where the linearization points are chosen on the boundaries of the regions, would lead to an upper bound on the probability of interest. Thus, in this case, optimal linearization points can be found by minimizing:

$$P(\ell_1 < 0 \cap \ell_2 < 0)$$

with respect to the linearization points, assuming that these are chosen from the set of boundary points to the intersection of regions (15.16).

5 FINAL REMARKS

The paper provides a lot of good ideas about how to apply FORM approximation to the problem of calculating out-crossing probabilities. The extremeley general setup allows us to analyse virtually any kind of problem. However, the actual precision of the answer may be questionable. In certain cases it is probably quite acceptable. In other cases the results may be less useful. Thus, when the methods are used, it is very important to fully understand the limitations of the FORM approximations.

An interesting feature of the proposed methods is that they allow us to incorporate physical knowledge (e.g., in terms of a set of differential equations) into the assessments. To apply these methods to real life situations, however, a lot of work needs to be done in order to develop sensible probability models for the uncertain quantities in the physical model.

16

REPLY TO THE REVIEWS

Armen Der Kiureghian, and Chun-Ching Li

Department of Civil Engineering,
University of California, Berkeley, California 94720

1 INTRODUCTION

The authors greatly appreciate the review comments and constructive criticism by T. Vrouwenvelder, A. B. Huseby and A. W. Kraak. It has been a privilege to have these prominent engineers and researchers read the paper so carefully and prepare detailed comments that have indeed helped us improve our method and extend its scope of application. We have learned that a method that was originally developed for application to earthquake engineering problems is potentially useful in assessing the reliability of flood defense systems. New formulations in this regard are presented at the end of this response.

2 DISCUSSION OF REVIEWERS' COMMENTS

Perhaps Vrouwenvelder's "engineer" rightfully characterizes the paper as "not understandable" and "scientific humbug." However, the truth is that this paper was the first attempt by the authors to explain their rather unorthodox approach, which makes use of the well known first-order reliability method (FORM) to transform classical random vibration problems into constrained optimization problems. As a result, the paper contains details that are unnecessary for the first reading and which portray the approach as more complicated than it actually is. As we gain experience, we find simpler ways of articulating our method that hopefully will make it more understandable to the "engineer."

R. Cooke et al. (eds.), Engineering Probabilistic Design and Maintenance for Flood Protection, 173–177.

Many engineering innovations go through this evolutionary process of gradual clarification and simplification, provided they prove to be useful in practice.

As late Professor Newmark of the University of Illinois would say, "an engineer has no right to solve a problem of which he/she does not already know the answer." The present authors are firm believers in the necessity for an engineer to have a feel for his/her problem and make rough estimates of the solution with simple methods of analysis. However, it is also wrong to think that all problems can be solved by simplified methods and that there is no need for sophisticated and mathematical algorithms that can address general classes of complex problems. Many advances in engineering would not have come about if we were to solely rely on simplified methods.

Vrouwenvelder shows his superb engineering skill by constructing a simple and accurate linearized solution for the duffing oscillator. However, it is not clear that he can do the same for more complicated systems, such as multi-degree-of-freedom, inelastic and hysteretic structures that are so common in earthquake engineering. The proposed method is meant to solve such general problems for which no effective methods of solution are currently available.

In his detailed review, Huseby formulates a number of problems that could be of interest in designing flood protection systems and wonders if the proposed method can solve them. In the last section of this response, we show that indeed this can be done. First, however, we would like to respond to two comments made by him. The first comment has to do with his equations (5)-(10) and the assertion that our proposed method requires an assumption of independence of out-crossing events from prior out-crossings and that it cannot account for system changes due to out-crossings. We disagree with this assertion. It is noted that any change in the characteristics of the system can and should be described through the differential equations governing the system behavior and response. This may include, for example, a change in the system properties (e.g., stiffness) when crossing of a certain threshold (e.g., yield level) occurs. A good example is the hysteretic oscillator discussed during the oral presentation by the first author, which is included in the expanded version of the paper. Furthermore, we note that $\nu(t)$ is a mean value that is computed from the marginal distribution of up-crossings, hence requiring no independence assumption.

Huesby's second comment has to do with the limitations of FORM. He states (paragraph following (18)):

> To make these work, it is implicitly assumed that the origin does not belong to any of the regions (16). Moreover, one need to assume that no region is a subset of the other. The main idea behind the methods is that the linear approximation should be good in areas with high probability, i.e., areas close to the origin.

We believe there is a misunderstanding here. As mentioned in item (4) of section 2 in the author's paper, the best linearization point is the one nearest to the point of maximum probability density *within domain* F, not within the entire space, i.e., the origin. Therefore, there is no restriction on the location of the origin. Moreover, for the up-crossing formulation presented, the two regions defined by (12) and (13) of the paper cannot be subsets of one-another.

As is well known, the accuracy of FORM depends on how well the linearized limit-state surface describes the domain F. While a precise measure of error cannot be given, often it is possible to determine if the approximation will work or not. In the present case, we note that when the system is deterministic and linear and the input is stochastic, then the response is a linear function of the input random variables $\mathbf{f}$ and the FORM approximation is exact if $G(\mathbf{u}(\mathbf{t}), \mathbf{v})$ is a linear function of $\mathbf{u}(\mathbf{t})$ (as is the case for threshold crossings of a scalar response process $u(t)$). Based on this observation, it is expected that FORM will provide a good approximation for nonlinear systems with deterministic properties. When the system properties $\mathbf{v}$ are also random, the accuracy then depends on the degree of nonlinearity of $G(\mathbf{u}, \mathbf{v})$ with respect to $\mathbf{u}$ and $\mathbf{v}$. If in doubt, benchmark studies with Monte Carlo simulation will have to be performed to ascertain the accuracy of the FORM approximation.

In his comments from the viewpoint of a "problem owner," Kraak describes a stochastic dynamic problem for jack-up platforms with uncertain properties and asks how the proposed method in conjunction with finite element codes (owned by oil companies) can be used to solve the problem. We believe that the proposed method can be very effective for such problems. Essentially, it would determine (through the optimization algorithm described in the paper) a sequence of deterministic systems and excitations for which the company-owned FE code should be run in order to make a FORM approximation of the probability of interest. The authors would gladly accept an opportunity provided by a "problem owner" to demonstrate this fact!

Kraak's second question is whether we can use the method to estimate the duration of exceedance of a threshold, e.g., over-topping of a dyke, by the proposed method. The answer is yes, as described in the following section.

3 HUSEBY'S PROBLEMS

In his discussion, Huseby elegantly formulates three problems of possible interest to design of flood defense systems. In this section, we show that these problems indeed can be solved by the proposed method.

Huseby's first problem deals with the time spent in the unsafe state during a given period $(0, T)$. He formulates it as:

$$\eta(T) = \int_0^T I[G(\mathbf{u(t)}, \mathbf{v}) \leq \mathbf{0}]\, \mathbf{dt},$$

where $I[E]$ is an indicator variable such that $I[E] = 1$ if E is true and $I[E] = 0$ if E is false. It is easy to show that the mean and mean square of $\eta(T)$ are given by

$$E[\eta(T)] = \int_0^T P[G(\mathbf{u(t)}, \mathbf{v}) \leq \mathbf{0}]\, \mathbf{dt}, \tag{16.1}$$

$$E[\eta^2(T)] = \int_0^T \int_0^T P[G(\mathbf{u(t_1)}, \mathbf{v}) \leq \mathbf{0} \cap \mathbf{G}(\mathbf{u(t_2)}, \mathbf{v}) \leq \mathbf{0}]\, \mathbf{dt_1}\, \mathbf{dt_2}. \tag{16.2}$$

The probability terms in both integrals can be easily computed by the proposed FORM method. We note in particular that $P[G(\mathbf{u(t)}, \mathbf{v}) \leq \mathbf{0}]$ requires a component reliability analysis for each t and $P[G(\mathbf{u(t_1)}, \mathbf{v}) \leq \mathbf{0} \cap \mathbf{G}(\mathbf{u(t_2)}, \mathbf{v}) \leq \mathbf{0}]$ requires a parallel system reliability analysis for each pair of t_1 and t_2. Interestingly, for a linear system and when $G(\mathbf{u}, \mathbf{v})$ is a linear function of $\mathbf{u}$ and $\mathbf{v}$, these FORM solutions are exact. Having computed the probabilities for each t_1 and t_2, the mean and mean square of $\eta(T)$ are computed from Eq. 16.1 and Eq. 16.2 by numerical integration. The variance can then be computed as $V[\eta(t)] = E[\eta^2(t)] - E^2[\eta(t)]$.

Huseby's second problem is defined by

$$A(T) = \int_0^T \max\{-G(\mathbf{u(t)}, \mathbf{v}), 0\}\, \mathbf{dt}.$$

If we consider the plot of $G(\mathbf{u(t)}, \mathbf{v})$ as a function of time, $A(T)$ represents the area below the zero level as shown in Figure 1. This could be related, for example, to the accumulated damage in a structure, or the amount of water released by over-topping of a dyke. It is easy to derive the following expression for the expected value of $A(T)$:

$$E[A(T)] = \int_0^T \int_0^\infty \theta \, \frac{\partial P[G(\mathbf{u(t)}, \mathbf{v}) + \theta \leq \mathbf{0}]}{\partial \theta}\, d\theta\, dt. \tag{16.3}$$

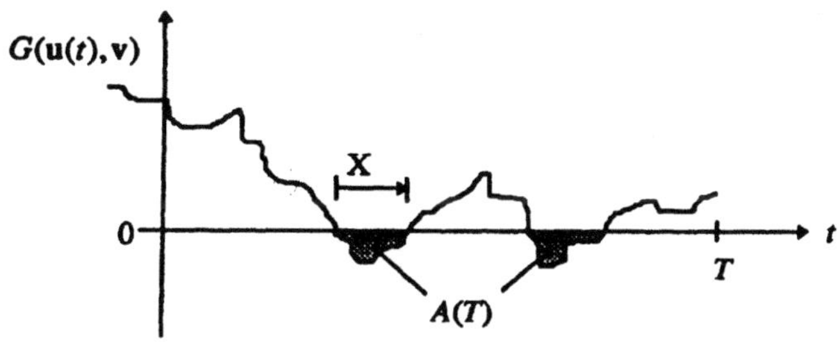

Figure 1 Definition of $A(T)$ and X.

The partial derivative in Eq. 16.3 is the sensitivity of $P[G(\mathbf{u}(\mathbf{t}), \mathbf{v}) + \theta \leq \mathbf{0}]$ with respect to the parameter θ. This sensitivity is easily computed by FORM. Hence, the mean value of $A(T)$ can be computed by the proposed method through a series of repeated FORM analysis for a range of θ and t values followed by numerical integration of Eq. 16.3.

Huseby's third problem is defined by

$$X = \inf\{\tau - t > 0 : G(\mathbf{u}(\tau), \mathbf{v}) > \mathbf{0}\},$$

where t is the time of an out-crossing. Clearly, $X(t)$ denotes the duration that the system stays in the unsafe domain following an out-crossing (e.g., the duration of a flood), as shown in Figure 1. If the processes involved are stationary, a simple approximation of the mean of $X(t)$ is

$$E[X(t)] \cong \frac{E[\eta(t, \Delta t)]}{\nu t \Delta t},$$

where $\nu(t, \Delta t)$ is the occupancy time during the interval $(t, t + \Delta t)$ and Δt is a small time interval. One can easily show that the above is equivalent to

$$E[X(t)] \cong \frac{P[G(\mathbf{u}(\mathbf{t}), \mathbf{v}) \leq \mathbf{0}]}{\nu(\mathbf{t})}$$

Both terms on the right-hand side are easily computed by the proposed method, as described earlier.

In closing, we would like to once again express our sincere thanks to the discussants for their valuable contributions, as well as the organizers of the conference for making this dialog possible.

PART IV

17

PROBABILISTIC DESIGN OF BERM BREAKWATERS

Han Vrijling, and Pieter van Gelder

Department of Civil Engineering,
Delft University of Technology, Delft, The Netherlands

ABSTRACT

In mechanical engineering the approach of optimal maintenance is mostly based on the empirical failure rate. Due to the unique character of hydraulic engineering structures, empirical failure data are lacking and a fundamental approach is needed. For the modelling of the aging process the serviceability limit state models are applied to the initial resistance of the structure taking account of the uncertainties involved. The probability of failure as a function of time is calculated by confronting the diminishing resistance with the loading. Within this model the usefulness of optimal repair intervals will be shown for a berm breakwater. In the last decade a renewed interest is shown for the berm breakwater concept. In the design of a berm breakwater, many limit states have to be checked by the designer: the stability of the armour cross and longshore, the compliance of the subsequent layers with filter rules, the stability of the toe under wave action, the geotechnical stability, overtopping, etc. In this paper attention will be focussed on the stability of the outer armour. Special attention is given to the longshore transport case. It will be shown that only a probabilistic approach gives insight into the sensitivity of this failure mechanism.

1 INTRODUCTION

In order to protect coastal lines of defence from being damaged by severe hydraulic loadings from the sea, berm breakwaters can be constructed. The characteristic feature of berm breakwaters is that the originally built dynamically stable profile becomes statically stable after some period of time in which "wind and waves have acted as contractor". However, oblique wave attack can initiate longshore transport of stones along the center line of the berm breakwater. To

R. Cooke et al. (eds.), Engineering Probabilistic Design and Maintenance for Flood Protection, 181–198.

avoid failure due to severe longshore transport, berm breakwaters have to be inspected and, if necessary, to be repaired.

In this paper we will develop a model for the failure behaviour and optimal repair maintenance strategy of the outer armour of a berm breakwater in a monsoon - hurricane wave climate. The paper is organised as follows: The meaning of a dynamically stable armour layer will be explained in Sec. 2. A model for the longshore transport of the armour units is given in Sec. 3. The berm breakwater is exposed to the wave climate described in Sec. 4. Three probabilistic approaches will be presented to describe the longshore transport probabilistically in Sec. 5. Not only the probability of failure of the outer armour layer at the end of its lifetime will be of interest, but also the failure behaviour during lifetime. This will be the subject of Sec. 6. Finally we present the maintenance decision model for minimising the costs of repair and failure in Sec. 7. The conclusions and summary of notation are given in Sections 8 and 9.

2 THE DYNAMICALLY STABLE ARMOUR LAYER

The most important function of a breakwater is to break waves coming from deep water towards the coast. The breaking of waves reduces the energy considerably and therefore prevent the coast from being affected by severe hydraulic loadings. Two types of breakwaters can be discerned: statically stable - and dynamically stable breakwaters. Statically stable breakwaters are breakwaters where no or minor damage is allowed under severe hydraulic loadings. Damage is defined as displacements of armour units. The mass of individual units must be large enough to withstand severe wave forces. Dynamically stable breakwaters are structures where profile development is concerned. Material around the mean sea level is continuously moving during each run-up and run-down of waves, but when the net transport has become zero the profile has reached its equilibrium. In a berm breakwater the stones are dynamically stable. In the case of a severe storm, the material is redistributed by the wave attack to form a natural profile as shown in figure 1 [4].

The design of the armour of a berm breakwater was brought on a more scientific footing by Van der Meer [5]. He developed a computational model to predict the cross-sectional profile of the armour layer after the reshaping process. The variables gouverning the reshaping process and the final profile are:

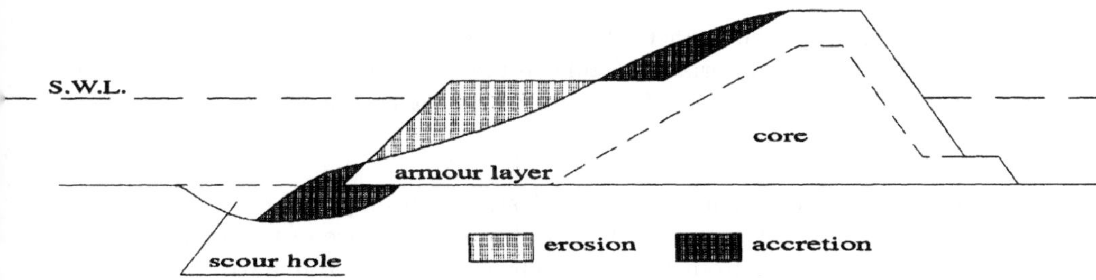

Figure 1 Cross sectional profile of a berm breakwater (SWL: still water level).

1. D_{50}, the size of the armour units (in metres);
2. Δ, the relative mass density (dimensionless) of the armour stone in water;
3. H_s, the significant wave height (in metres);
4. T_p, the spectral peak wave period (in seconds);
5. N, the number of waves (dimensionless);
6. h, the water depth in front of the structure (in metres).

A rough indication of the stability of the armour is given by the ratio // $H_0 = H_s/D_{50}$. According to [5] this ratio should lie in the range 3 - 6 for berm breakwaters. It should however be noted that in two-dimensional flume tests structures with higher H_0 ratio's can be perfectly stable under wave attack after the reshaping process has taken place. Sand dunes where this ratio exceeds the value of 300 provide an example. In such cases the longshore transport of material quickly gains importance. Two dimensional model tests on berm breakwaters confirm this hypothesis. It is very difficult to destroy a model berm breakwater by incoming waves, unless the structure is overtopped. Overtopping waves easily damage the back slope of the breakwater. Consequently an erosion process may start that quickly causes a breach in the breakwater.

3 LONGSHORE TRANSPORT OF THE ARMOUR LAYER

The reshaping process of the cross-section places no obvious upper bound on the H_0 ratio. However for high values of this ratio a considerable longshore transport will take place when the waves attack under an angle. In cases where the amount of material available on the windward side is insufficient the breakwater will collapse. The longshore transport case is not well researched for berm breakwaters.

In [2] a first contribution is made by testing a reshaping breakwater under three varying angles of wave attack. However no influence of the angle of wave attack can be deduced from the data. On the basis of Burcharth's data and a berm breakwater design of Vrijling et.al. [6], a model of the following form is proposed to predict the longshore transport after the reshaping process has taken place:

$$S = A[(H_0T_0 - B)^C]_+ \tag{17.1}$$

where $[x]_+ = \max\{0, x\}$,
$H_0 = H_s/D_{50}\Delta$ and $T_0 = T_p\sqrt{g}/D_{50}$.
A, B and C are dimensionless constants based on empirical values 4.8×10^{-5}, 100.0, 2.0, and g is the acceleration due to gravity 9.8 m/s^2. S is the longshore transport of number of stones per wave of primary armour material passing through a certain cross-section of a berm breakwater.

4 THE WAVE CLIMATE

We will analyze the longshore transport in a monsoon - hurricane wave climate. The significant wave height (defined as the mean of the one third heighest waves) of the monsoon waves, H_m, are modelled by a Gumbel distribution ($F_{H_m} \sim$ Gumbel) with a mean of 2.10 m and a standard deviation of 0.36 m [6]. The wave steepness in a monsoon wave field, s_m, is also modelled by a $F_{s_m} \sim$ Gumbel. We assume the wave steepness to be independent of the wave height.

The significant wave height of the hurricane waves, H_h, are modelled by a Frechet distribution $F_{H_h} \sim$ Frechet with a mean of 3.19 m and a standard deviation of 1.80 m. The wave steepness, s_h, in this wave field is modelled with a normal distribution $F_{s_h} \sim$ Normal (also independent of the wave height).

The wave height just in front of the breakwater is limited to $h = 5.0$ m by the water depth. We assume the monsoon to take place every year for a period of 3 months. The wave heights during these 3 months are modelled by 3-hourly independent realisations from the Gumbel distribution. In a section of 3 hours the wave height is assumed to be constant. The number of waves in the planned lifetime of the structure of 50 years in the monsoon wave field is given by: 50 years x 3 months x 30 days x 8 sections x $N_3 = 36000\ N_3$. Here is N_3 the number of waves in a section of 3 hours and is given by 3 hours x 3600 seconds / T_m, where the wave period in the monsoon wave field, T_m, is calculated by $T_m = \sqrt{H_m/1.56 s_m}$. So note that N_3 is a function of H_m and s_m.

In the hurricane wave field the number of waves is given by 50 years x N_{24}, where N_{24} is the number of waves in a section of 24 hours and is given by: 24 hours x 3600 seconds / T_h, where T_h is the wave period in the hurricane wave field and calculated by $T_h = \sqrt{H_h/1.56 s_h}$.

5 THREE PROBABILISTIC APPROACHES

5.1 Probabilistic approach; a first step

The expected value of the longshore transport during the lifetime of a berm breakwater in the wave climate of Sec. 4 can be derived as:

$$\begin{aligned} E(S_{life}) &= 36000 \int_0^{H_{m_{max}}} \int_0^{s_{m_{max}}} N_3 S f_{H_m}(H) f_{s_m}(s) dH ds + \\ &+ 50 \int_0^{H_{h_{max}}} \int_0^{s_{h_{max}}} N_{24} S f_{H_h}(H) f_{s_h}(s) dH ds, \end{aligned} \tag{17.2}$$

where $H_{m_{max}}$ and $H_{h_{max}}$ are determined using extreme value theory: $F_{H_{m_{max}}}(x) = F_{H_m}^{N_m}(x)$, where N_m the number of 3-hourly independent realisations in the monsoon wave climate during the planned life time = 36000. $FH_{h_{max}}(x) = F_{H_h}^{N_h}(x)$, where $N_h = 50$, because we assume that only one hurricane occurs in each year. Further $s_{m_{max}}$ and $s_{h_{max}}$ are both assumed to be 0.05. Waves can't become steeper.

The expected value of the number of stones transported during the monsoon over the entire life of the structure can be calculated by substituting in the first part of integral (17.2), the p.d.f. of the wave height and the wave steepness related to the Gumbel distribution. If the integration is performed with the transport relation (17.1), the expected value of the transport is 258 stones.

The expected value $E(S_{life})$ of the transport is also dependent on uncertain values as A and B in the transport relation. If for the coefficients in the transport relation the following safe values are chosen $A = 7.2 \times 10^{-5}$; $B = 80$, the expected value of the transport for the Gumbel distributed wave height in a monsoon wave field becomes 2412 stones or 2412 x $M_{50} = 9000$ ton. A similar approach is followed for the hurricane wave climate by substituting in the second part of integral (17.2) the Frechet p.d.f. for the wave height and the Normal p.d.f. for the steepness. Also in this case the transport is sensitive for changes of A and B:

average transport relation	"safe" transport relation
499	2923

As the transport by the monsoon and by the hurricane have to be added to find the amount of nourishment at the weatherside, the following values should form the basis of the design.

Total transport

average transport relation	"safe" transport relation
757	5335

The designer is confronted with a difficult decision.

5.2 Probabilistic approach; a second step

The correct solution for the posed above is found by introducing the uncertainties described into the design equations in the form of additional p.d.f.'s. Moreover the uncertainty of the Δ and D_{50} values (hidden in the dimensionless quantity H_0T_0) that will be realised during construction and the exact amount of nourishment that will be supplied can be included.

By means of a Level II probabilistic calculation the influence of the uncertainty in D_{50}, the coefficients A, B in the transport relation and the integration bounds $H_{m_{max}}$ and $H_{h_{max}}$ on the expected transport can be calculated. And if the amount of material at the windward side to nourish the longshore transport is known the probability of failure can be established. To this end the following reliability function is defined:

$$Z = R_0 - S_{life}, \tag{17.3}$$

where R_0 stands for the nourishment.

We assume the parameters, A, B, Δ, D_{50} to be normally distributed. The nourishment R_0 is also normally distributed. The level II calculation that is performed shows in Table 1 that the probability of failure is 29% in the life time of the breakwater.

Table 1 Result of a Level II probabilistic calculation

Name	Type	A	B	C	μ	σ	X	α^2
Δ	N	0	0	0	1.570	0.060	1.564	0.037
D_{50}	N	0	0	0	1.150	0.058	1.138	0.143
A	N	0	0	0	4.8×10^{-5}	2.4×10^{-5}	0.000	0.062
B	N	0	0	0	100	20	90.43	0.757
$H_{m_{max}}$	G	4.920	0.284	0	5.024	0.372	5.024	0.000
$H_{h_{max}}$	F	0	7.926	3.388	8.832	3.001	8.832	0.000
R_0	N	0	0	0	2500	200	2495	0.002

$Z(X) = -0.2689$ (X is the design point), $\beta = 0.550$.

Failure implies here that the extra amount of nourishment provided at the time of construction is completely taken away during the design life.

The level II calculation also shows the contribution of the variables to the standard deviation of Z in the column α^2. The values in the column indicate that the greatest contribution comes from the B coefficient in the transport relation, with the D_{50} in the second place. The integration boundaries, formed by the maximal significant wave heights due to the monsoon and the hurricane during the life time are relatively unimportant.

From the table the conclusion may be drawn that further research should be done to establish the transport relation, especially the start of the movement modelled by the coefficient B. During construction the D_{50} of the rock going into the berm should be strictly controlled by quality assurance measures to prevent errors to the lower side.

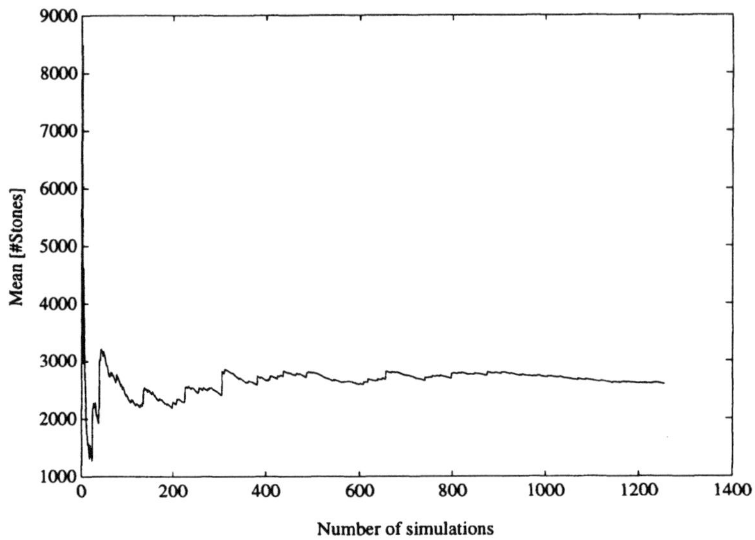

Figure 2 Influence of the number of MC-simulations on $E(S_{life})$.

5.3 Probabilistic approach; a third step

The next step is to perform a Level III analysis, which means a Riemann integration of the following integral:

$$E(S_{life}) = \int \dots \int NSf_A(a)f_B(b)f_\Delta(\delta)f_{D_{50}}(D_{50})f_H(H)f_s(s)dadbd\Delta dD_{50}dHds \tag{17.4}$$

Riemann integration of this integral turns out to be almost impossible due to extensive calculation time. We therefore apply a Monte Carlo simulation. As input we have normal distributions for A, B, D_{50}, Δ with the same means and variances as given in Table 1. We have performed the MC simulation with 500 realisations at 50 years lifetime. With this number of realisations we have converged to a steady value of $E(S_{life})$. See figure 2.

The 500 simulated values of S_{life} are plotted in sorted order in figure 3.

An immediate conclusion is that there is a very large spreading in the expected number of stones in the longshore transport. The mean value of S_{life} is given by 2449 stones. The standard deviation is 6786 stones (!). It is also seen from the picture that there is about 8% chance that there are no stones transported during the breakwater's lifetime at all. With a statististical fit program of [3], it is derived that the lognormal distribution function fits the data of S_{life} best. Let the lognormal fit for S_{life} be denoted by $F_{S_{life}}$. If the probability density function of the nourishment is denoted by f_R, then failure of the berm breakwater during its life time, $p_{f_{life}}$ is given by:

$$p_{f_{life}} = \int (1 - F_{S_{life}}(x)) f_R(x) dx \qquad (17.5)$$

If we assume the nourishment to be modelled by a normal distribution with the same mean and variance as given in Table 1, then $p_{f_{life}}$ can be calculated to be 0.23. In comparison with the rate of 0.29 in the level II calculation, it seems that the level II calculation was a bit too conservative.

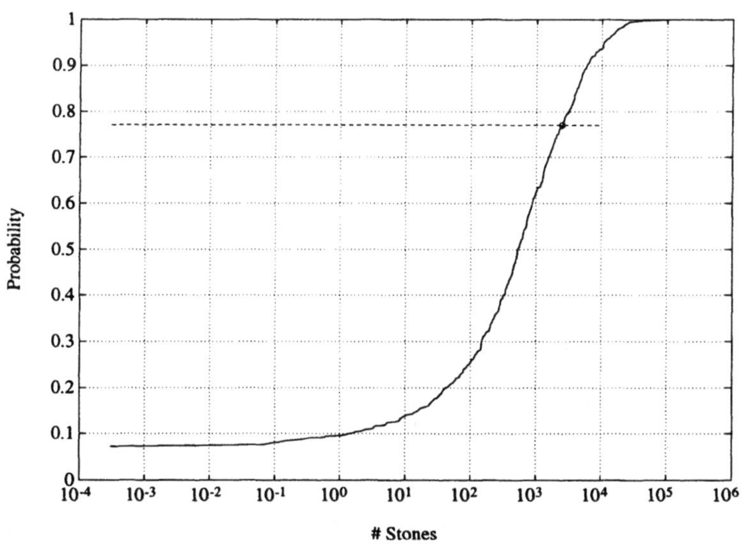

Figure 3 Simulated values of S_{life} including its mean (o) and standard deviation (dotted line).

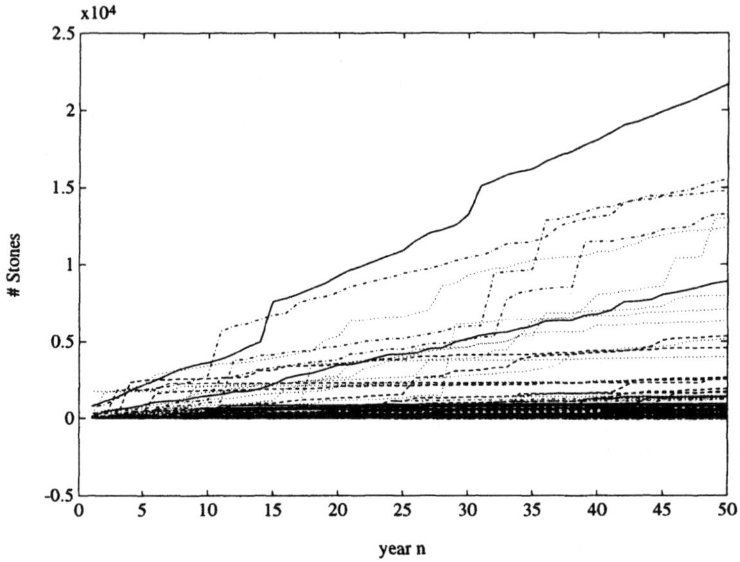

Figure 4 The number of transported stones until and including year n for 100 simulations.

6 FAILURE BEHAVIOUR DURING LIFETIME

We will now analyze the behaviour of the probability of failure of the breakwater as a function of its age. We perform a Monte Carlo simulation in which we follow the breakwater during its 50 years concerning the amount of transported stones. The simulation is being repeated 500 times and the results for the first 100 realisations are shown in figure 4.

The stepwise behaviour of some of the graphs is caused by the occurence of hurricanes with extreme wave heights by which many stones are transported.

For every year n, we are interested in the probability distribution of the amount of transported stones until that year. Our Monte Carlo simulation gives figure 5 for this.

In figure 5, the distribution functions for every 5 years are plotted. For n to infinity, the distribution will tend to the normal distribution function. For n=50, we are however, as we saw before, still in the lognormal situation. From the above figure we can also derive the probability function of No-Transport of stones until year n. In the first year there is a chance of almost 50% that there is no transport at all. This chance tends to 0 as the life of the breakwater progresses (see fig. 6).

We can calculate the probability of failure of the breakwater during its lifetime. We study two different sizes for the mean of the normally distributed nourishment in the first year: (i) 2500 stones and (ii) 5000 stones. With the Monte Carlo simulation we obtain a progress of the probability of failure as shown in fig. 7.

With the nourishment two times larger, the probability of failure reduces more or less 2 times during its entire life.

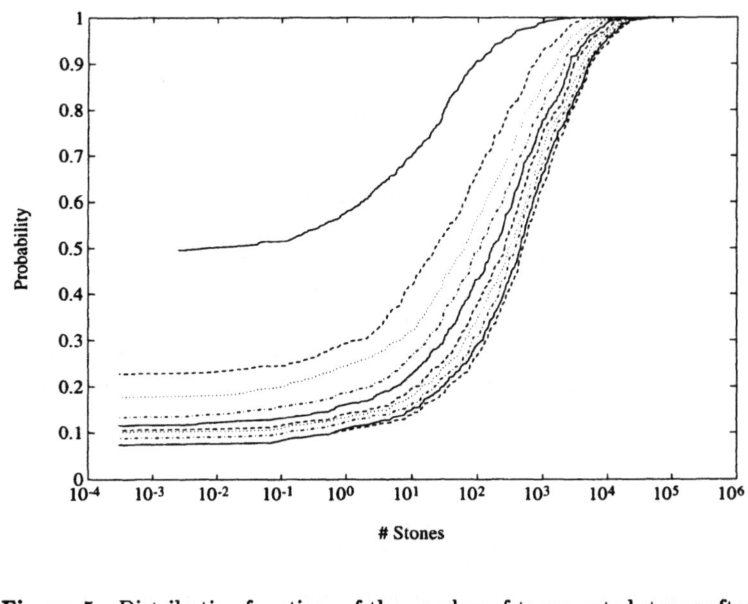

Figure 5 Distribution functions of the number of transported stones after n years (n=1,5,10,..,50).

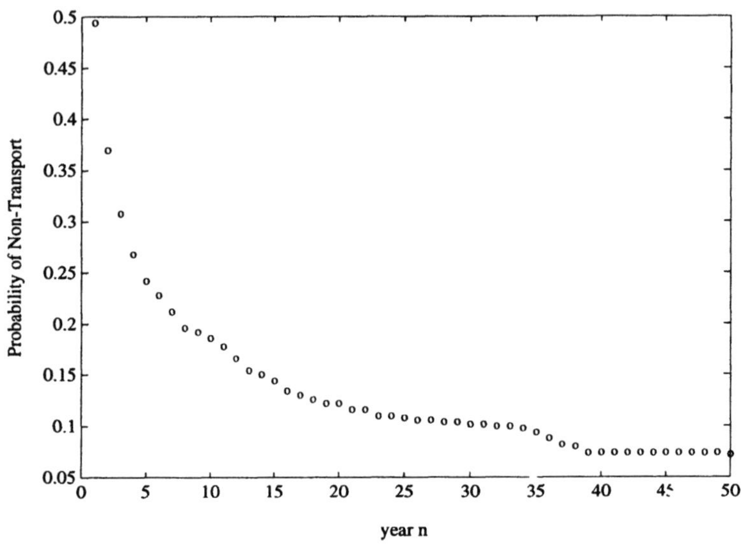

Figure 6 Probability of non-transport as a function of year n.

7 A MAINTENANCE MODEL

We are now in the position to apply some maintenance theory on our berm breakwater. We focus on one section of the breakwater where the nourishment is applied. We are interested in determining the optimal interval for repairing the berm breakwater; i.e. bringing the breakwater back to the original state. The present value of the expected costs of failure of the berm breakwater can be written as follows:

$$\sum_{n=1}^{50} c_f p_n \frac{1}{(1+r)^n}, \tag{17.6}$$

where: c_f are the costs of failure (200,000 guilders), consisting of loss of assets facilities and loss of income due to the failed breakwater. r is the discount rate (0.04). p_n is the probability of failure in year n.

Let S_i be the number of transported stones in year i, i=1,2,...,50. We assume that the cost of repair in year n is proportional to the expected erosion in year n, $E(\sum_{i=1}^{n} S_i)$. The present value of the total cost of repair in year n is given by:

$$\frac{c_m + c_v E(\sum_{i=1}^{n} S_i)}{(1+r)^n}. \qquad (17.7)$$

where: c_m are the mobilization costs for one repair (10,000 guilders) and c_v are the costs per stone (100 guilders). Notice that $E(\sum_{i=1}^{n} S_i) = nE(S_1)$.

Now the present value of the costs of repairs have to be added over the life time of the structure. This defines the objective function that has to be minimized with respect to ΔT, the repair interval (in years).

$$\begin{aligned} K(\Delta T) \quad = \quad & \sum_{j=1}^{DIV(50,\Delta T)} \frac{c_m + c_v j \Delta T E(S_1)}{(1+r)^{j\Delta T}} + \qquad (17.8) \\ & + \sum_{j=1}^{DIV(50,\Delta T)} c_f \sum_{n=1}^{\Delta T} \frac{1}{(1+r)^{(j-1)\Delta T + n}} \sum_{i=n}^{\Delta T} \frac{1}{(1+r)^i} p_n + \\ & + \sum_{n=1}^{MOD(50,\Delta T)} c_f \frac{1}{(1+r)^{50-MOD(50,\Delta T)+n}} \sum_{i=n}^{MOD(50,\Delta T)} \frac{1}{(1+r)^i} p_n . \end{aligned}$$

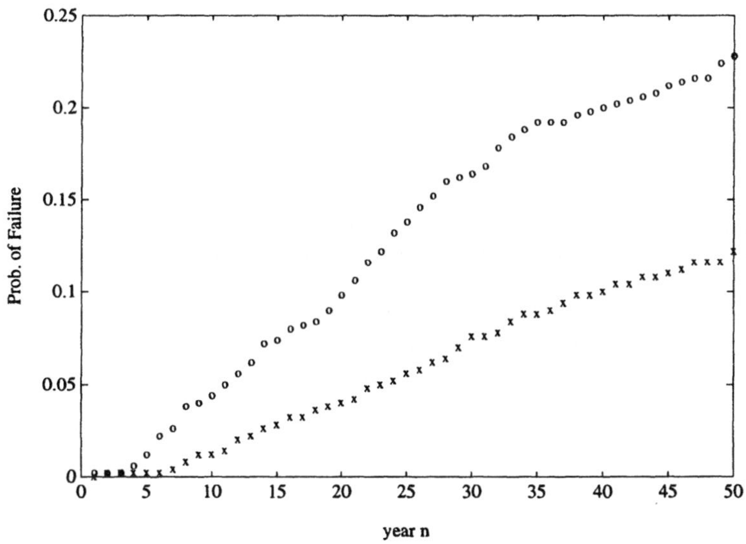

Figure 7 Cumulative probability of failure as a function of year n, for $\mu(R_0) = 2500$ (o) and $\mu(R_0) = 5000$ stones (x).

$DIV(50, \Delta T)$ denotes the number of repairs in 50 years where the repair interval is ΔT years. $MOD(50, \Delta T)$ denotes the number of years between the last repair and the end of the lifetime of the breakwater at 50 years. Remark that $DIV(50, \Delta T) * \Delta T + MOD(50, \Delta T) = 50$.

The present value of the total cost $K(\Delta T)$ as a function of ΔT is given in fig. 8.

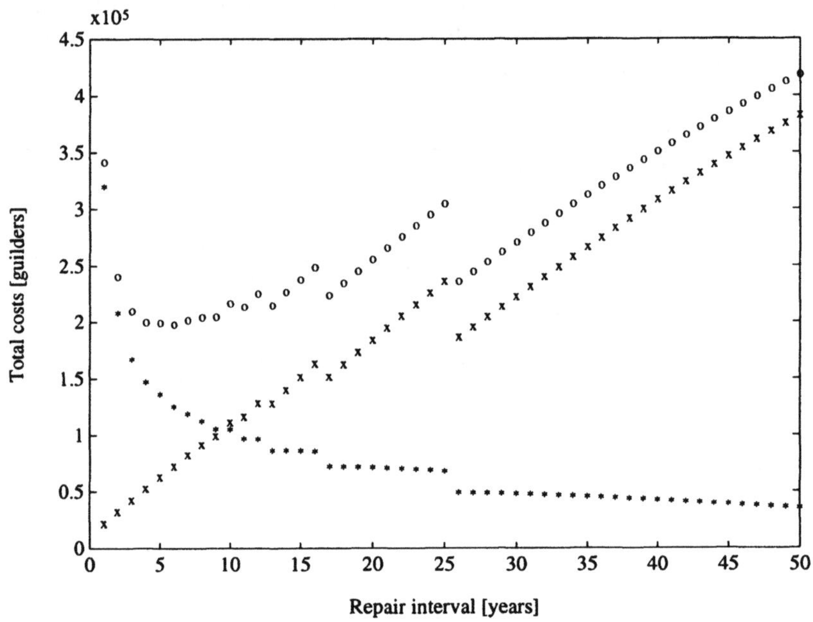

Figure 8 The expected costs (o), the expected costs of repair (*) and the expected costs of failure (x) over a bounded time horizon of 50 years.

The optimal repair interval appears to be 6 years. We can analyze the sensitivity of $E(S_1)$ on the optimal repair interval by substituting $E(S_1)$ by $E(S_1) + \alpha\sigma(S_1)$ (the probability of failure changed accordingly). We obtain table 2.

Table 2 α **versus** ΔT

α	ΔT
0	6
$\frac{1}{2}$	4
1	3
$1\frac{1}{2}$	2

We observe a large sensitivity. The more uncertainty in the rock displacement process, the smaller the repair interval.

8 CONCLUSIONS

The paper shows that although the design of the cross section of a berm breakwater is well defined by the work of Baird [1], Van der Meer [5] and others, the longshore transport caused by angular wave attack may pose a threat that must be addressed by the designer and the researcher of this type of structures.

A first step towards probabilistic approach, that takes into account the variability of the waves, shows that the expected value of the longshore transport has a positive value and that a nourishment has to be provided. It is shown that the decision on the amount of nourishment is far from simple in the light of the uncertainties in the transport relation. The amount may vary from 757 stones to 5335 stones depending on the optimism of the designer.

In a second step all uncertainties are included in a level II probabilistic calculation. It is shown that a nourishment of 2500 stones has a probability of failure of 29% in the life time. The level II calculation shows further that the uncertainty of the transport relation contributes most to the failure probability. So more research into this problem is needed. Secondly the stone size has to be carefully controlled during construction by quality assurance procedures.

In a third step a level III probabilistic calculation is performed. With the same parameters as in the level II calculation a slightly smaller probability of failure is found: 23% in the lifetime. The number of transported stones are succesfully modelled by a lognormal distribution function. The variance of the number of transported stones is much higher as its mean value. This is being caused by the extreme values of the wavefield in a hurricane.

The level III approach has the advantage on the level II approach that it is an exact calculation (within the framework of the model) of the probability of failure and that it gives us the variance of the parameters of interest as well. The calculation time in the Level II approach is however much shorter and besides it gives us a quick insight of the influence of the model parameters on the probability of failure.

The probability of failure of the bermbreakwater is analyzed as a function of time. A Monte Carlo simulation method has been used for this. With this result a maintenance model of the berm breakwater is analyzed. The optimal repair interval turns out to be 6 years, but very sensitive for uncertainty in other parameters.

9 SUMMARY OF NOTATION

parameter	description	mean value	dimension
D_{50}	size of the armour units	1.15	m
Δ	relative mass density of the armour stone in water	1.57	kg/m^3
H_s	significant wave height		m
s_p	wave steepness		
T_p	spectral peak wave period		s
N	number of waves		
h	water depth in front of the structure	5	m
g	gravitional constant	9.81	m/s^2
H_0	stability indicator of the armour (given by the ratio H_s/D_{50})		
T_0	dimensionless wave period (given by $T_p\sqrt{g/D_{50}}$)		
S	number of stones in longshore transport per wave		
A, B, C	constants in transport relation		
H_m	significant wave height in monsoon wave field	2.10	m
H_h	significant wave height in hurricane wave field	3.19	m
s_m	wave steepness in monsoon wave field		
s_h	wave steepness in hurricane wave field		
S_{life}	number of stones in longshore transport per wave during life time (of 50 years)		
c_f	costs caused by a possible failure	200,000	guilders
r	discount rate	4%	
p_{life}	probability of failure in the lifetime of the breakwater		
p_n	probability of failure in year n		
S_i	number of stones of stones in longshore transport in year i		
c_m	mobilization costs for one repair	10,000	guilders
c_v	costs per stone	100	guilders
K	costs of maintenance during life time		guilders
ΔT	inspection interval		years

REFERENCES

[1] W.F. Baird and K.R. Hall. The design of armour systems for the protection of rubble mound breakwaters. In *Proc. Conf. on Breakwater design and construction*. Institute of Civil Engineering, London, 1983.

[2] H.F. Burcharth and P. Frigaard. Reshaping breakwaters, on the stability of roundheads and trunk erosion in oblique waves. In D.H. Willis, W.F. Baird, and Magoon O.T., editors, *Berm Breakwaters: Uncoventional Rubble-Mound Breakwaters*, pages 55–72. American Society of Civil Engineers, September 1987.

[3] M. Hauer. Chifit; fit programmature, February 1995.

[4] Saskia Plate. Probabilistic design of breakwaters: The ennore breakwater project, failure mechanisms of a crested rubble mound breakwater. Master's thesis, Delft University of Technology, The Netherlands, June 1995.

[5] J.W. Van der Meer. Application of computational model on berm breakwater design. In D.H. Willis, W.F. Baird, and Magoon O.T., editors, *Berm Breakwaters: Uncoventional Rubble-Mound Breakwaters*, pages 93–103. American Society of Civil Engineers, September 1987.

[6] J.K. Vrijling, E.S.P. Smit, and P.F. De Swart. Berm breakwater design - the longshore transport case: a probabilistic approach. In Thomas Telford, editor, *Coastal Structures and Breakwaters*, pages 403–415. Institute of Civil Engineering, London, 1991.

18

REVIEW PROBLEM OWNER PERSPECTIVE

Rommert Dekker

Econometric Institute, Erasmus University, Rotterdam, The Netherlands

The paper deals with the design and maintenance of berm breakwaters and focusses on the longshore transport in a monsoon-hurricane wave climate. The main object is to determine the initial nourishment needed to survive the design life of 50 years. Three probabilistic approaches are applied to the problem. In the user's perspective we do not necessarily focus on the correctness of the approach but on the appropriateness of the approach for the problem on hand. The user in this respect is the designer of a berm breakwater which has to guarantee a 50 year life and which has little means to get information during the life of the berm breakwater other than at its failure. This aspect is essential to understand the reasons behind the focus on the uncertainty in the long shore transport.

The three approaches provide an increasing sophistication in the way to tackle the problem. The first approach appears to be far too simple, the second is somewhat conservative. Finally, the third approach, although exact, has to be dealt with by simulation. They all yield that there is substantial uncertainty in the amount of transport.

There are several sources of uncertainty. First of all there is the climate with its monsoon and hurricane waves. Their height, steepness and angle all effect the long-shore transport. Secondly the transport equation (1) is not certain either. It contains a threshold B which plays an important role (see Table 1). Thirdly, the quality of the berm breakwater, through the average size of the armour units D_{50} and the relative density Δ, is important, though to a lesser extent.

R. Cooke et al. (eds.), Engineering Probabilistic Design and Maintenance for Flood Protection, 199–200.

The first source is outside our control and is inherently uncertain. Especially the hurricanes can have a devastating effect (cf fig. 4). It is strange that figure 4 suggests that the effect of the monsoons is linear per realisation over the lifetime. The second source is a matter not knowing. This could be handled by doing more research into the transport equation for practical situations, or alternatively, it is where alternative actions should take place to prevent the long shore transport in the design on hand. Finally, the quality of the breakwater is within direct control. The question which now comes forward, is what is the cheapest way to reduce the uncertainty. This question is deepened by incorporating the possibility of corrective maintenance as is done in section 7. Although the probability of failure within 50 years appears to be 0.29, it is economically optimal to repair the berm breakwater every 6 years! It would be interesting to know what would be cheaper: repairing every 6 years or having a larger initial nourishment.

There may be some critique on the maintenance policy. The authors use a periodic repair policy within a finite horizon. They could also have used a nonperiodic policy since the berm breakwater is initially in a good condition. Even better would be a policy where the timing of the repair would be state dependent (e.g. the amount left over of the initial nourishment). The latter policy would require a proper inspection every few years and when that is questionnable one has to take the aforemenentioned policies.

Another aspect to be discussed is how the results can be scaled up to a specific berm breakwater. The authors calculate the probability of failure for a specific cross section. The berm breakwater will have a certain size and it is not yet clear whether the failure probability concerns the whole breakwater or only specific parts.

19

REVIEW ENGINEERING PERSPECTIVE

Ad van der Toorn

Harbour of Rotterdam, Rotterdam, The Netherlands

1. Although figure 1 now gives insight in the reshaping-process of one cross-section of a berm-breakwater (is there really so much accreation at the top?), the really important figure about the longshore situation is missing. Which transport is predicted across which sections and what about the place of the nourishment? Is there a reserve in every cross-section?

2. The suggested 'correct solution' in chapter 5.2 is worked out only for a level II-approach with normally-distributed parameters. So fortunately there is still room for a better approach in 5.3!

3. There should be an explanation why the upper integration-bounds of the wave heights does have no influence on the outcome of Z (see table 1).

4. If the lesson from the level II-approach in 5.2 is that all of the uncertainty comes from the start of the movement modelled by coefficient B, then what is the reason to optimize the better part by a level III approach?

5. An engineer should check some points in figure 8, to get a good feeling about the optimum maintenance interval and to do so it allmost every time leads to 'new' figures! The first point should be a one-years-interval: There are no failure-cost, but only mobilization-costs 50 * Dfl 10.000 plus the cost of repair of the expected losses. Another point should be a ten-years interval: The probability of failure is about 0.05(?), so the risc should be less than 5 times 0.05*200.000= 50.000 (but that's not in fig.8 ?). The last point should be the 'design-point': T=6 years.

6. An engineer should start simply to check with r=0!

R. Cooke et al. (eds.), Engineering Probabilistic Design and Maintenance for Flood Protection, 201–202.

7. An engineer should optimize the total costs, so take into account the number of initial stones (2500, 5000, 7500, 10.000?).

8. A maintenance manager should change the D50 of the suppletion after one or two intervals, when he founds out that this is the main reason of his losses (see the cumulative effect in figure 4).

9. A maintenance manager should anticipate at the occurence of a hurricane which effect on the losses is equal to a six-years-period of monsoon, and so come to a more condition-based-maintenance.

20

REVIEW MATHEMATICAL PERSPECTIVE

Tom Mazzuchi

*Department of Engineering and Operations Research,
The George Washington University, Washington, DC 20052*

Vrijling and van Gelder (1995) provide a thorough first analysis of an interesting problem in civil engineering: berm breakwater design and maintenance. It is always easier to criticize than to construct, and I make my comments with this thought in mind. I hope I have provided some questions which, when addressed, will clarify their work and some suggestions which, if acted upon, will improve their analysis. I make my analysis section by section.

At the heart of any mathematical analysis is its assumptions. It is the assumptions upon which practitioners tend to focus when accepting or rejecting an analysis. Thus the assumptions should be clearly stated up front and motivated. This is the main fault I find in problem formulation sections. In the presentation of the problem definition, we find the following assumptions sprinkled throughout.

1. The planned lifetime of the breakwater is intended to be 50 years.
2. The longshore transport (number of stones) per wave is empirically given by

$$S = \begin{cases} A(H_0T_0 - B)^C, & \text{for } H_0T_0 > B, \\ 0, & \text{for } H_0T_0 \leq B, \end{cases}$$

   where $H_0 = H_S/\Delta D_{50}$, $T_0 = T_P\sqrt{g}/D_{50}$, H_S is the significant wave height or mean of the 1/3 highest waves), Δ is the relative mass density of the armor stone in water, D_{50} is the size of the armor T_P is the spectral peak wave period, g is acceleration due to gravity, and A, B, and C are constants empirically defined from data.

R. Cooke et al. (eds.), Engineering Probabilistic Design and Maintenance for Flood Protection, 203–207.

3. Significant longshore transport only occurs during monsoon and hurricane climates.
4. Monsoons take place every year for a period of 3 months.
5. During monsoons, significant wave heights remain approximately constant in a three hour period.
6. Significant wave height during a monsoon, H_m, is modeled by a Gumbel distribution with mean 2.10m and standard deviation 0.36m.
7. Wave steepness during a monsoon, s_m, is given by a Gumbel distribution.
8. Wave steepness and significant wave height in a monsoon are independent.
9. Over the lifetime of the breakwater it encounters the following number of waves due to a monsoon

$$50(\text{years}) * 3(\frac{\text{months}}{\text{year}}) * 30(\frac{\text{days}}{\text{month}})8(\frac{\text{3 hour sections}}{\text{day}}) * N_3(\frac{\text{waves}}{\text{3 hour section}})$$

where $N_3 = \frac{3(\text{hours}) * 3600(\frac{\text{seconds}}{\text{hour}})}{T_m}$ and $T_m = \frac{\sqrt{H_m}}{1.56 s_m}$

10. Hurricanes take place once a year for a period of 24 hours.
11. During hurricanes, significant wave heights remain approximately constant in a twenty four hour period.
12. Significant wave height during a hurricane, H_h is modeled by a Frechet distribution with mean 3.19m and standard deviation 1.8m.
13. Wave steepness during a hurricane, s_h, is given by a normal distribution.
14. Wave steepness and significant wave height in a hurricane, are independent.
15. Over the lifetime of the breakwater it encounters the following number of waves due to a hurricanes

$$50(\text{years}) * 1(\frac{\text{day}}{\text{year}})N_{24}(\frac{\text{waves}}{\text{hurricane day}})$$

where $N4_{24} = \frac{24(\frac{\text{hours}}{\text{day}}) * 3600(\frac{\text{seconds}}{\text{hour}})}{T_h}$ and $T_h = \frac{\sqrt{H_h}}{1.56 s_h}$.

16. Wave height just in front of the breakwater is limited to 5.0 m

All the assumptions seem to be standard modeling assumptions. However, they should be either motivated or referenced to works which have motivated them, as was done for assumption 2. There is some conflict between assumption 5 and 9. If separate three hour periods are used then why calculate the number of waves using the formula in 9, shouldn't it be the sum of all three hour realizations over the 50 year period? The same comment holds for 11 and 14. In addition, what is the basis for assumptions 4-8 and 10-14. Are there empirical studies to support these assumptions or are they made for mathematical convenience? Finally, what is the source for assumption 16 and why is this important to the analysis?

In the first step of the probabilistic approach, the authors calculate the expected longshore transport using the definition of longshore transport per wave times the number of waves occurring due to both monsoons and hurricanes. Both expressions are functions of significant wave height and wave steepness and thus the expectation is calculated using the probability distributions assumed for wave height and steepness for monsoons and hurricanes respectively. That is

$$E[S_{\text{life}}] = E[S_{\text{monsoons}}] + E[S_{\text{hurricanes}}].$$

Curiously the "expectation" is not taken over the full range of the random variables, but from 0 to some maximum value. These maximum values are calculated "using extreme value theory". In so doing the authors have not taken a proper expectation, neither with respect to the original distributions nor some fully defined conditionals. The use of extreme value theory in this situation is not at all clear.

In addition, the authors note that due to the definition of S, this expectation is a function of some constants A and B previously empirically specified. The authors then specify new values of A and B and recalculate the expectation. The authors refer to the first expectation as the "average" transport relation and the second expectation as the "safe" transport relations. Several questions arise. First, it is noted that S is a function of C as well as A and B. Why was sensitivity with respect to this value ignored. Second, where do the new values of A and B come from and why does the expectation with these values produce a "safe" transport relation. When taking uncertainty of estimated constants into account, it is common to use joint confidence regions. Is this what was done? If so, it should be clearly stated. If not, a motivation for concentrating on A and B and for specifying the particular values should be given.

In the second step of the probabilistic approach, the authors check for sensitivity to various other model constants by specifying normal distributions for these constants. This seems at best arbitrary but as noted in Lindley (1983) the elicitation and modulation of expert judgment concerning quantities of interest often involves the use of the normal distribution.

The second phase proceeds by what appears to be a regression analysis of the relation

$$Z = R_0 - S_{\text{life}},$$

though the constructed table is unintelligible. For clarity this section should be expanded with more detail; sensitivity analysis is an important part of any modeling phase. Curious is both the definition and interpretation of the above expression. It is termed the "reliability function " which is most often reserved for the complement of the cumulative distribution function. This may cause confusion with anyone familiar with reliability theory. In addition, the above function is used to denote failure when $Z = 0$. This is a simplifying assumption which may be too simplifying. Certainly it does not make sense for $Z = 1$ to be considered a success and $Z = 0$ a failure. Surely the success of the breakwater is a function of the magnitude of Z. This points one towards a loss function setup for Z rather than a straight difference. As the analysis is simulation based, this introduces only minor difficulties.

In the third step of the probabilistic approach, the authors take into account the uncertainty for all values by taking the expectation with respect to all uncertain quantities. Due to the complicated nature of the expectation equation, the authors resort to simulation. The authors do a good job motivating their simulation sample size based on the stability of $E[S_{\text{life}}]$. There is some confusion as to whether the authors are simulating a distribution for S_{life} or $E[S_{\text{life}}]$ as the graph caption indicates the former but when discussing the graph the authors refer to the "spreading in the expected number of stones in the longshore transport". Clearly the authors intended to develop a distribution for S_{life}.

The "best fit" for the simulated distribution for S_{life} is a lognormal distribution. The authors have not defined "best fit". Is the fit best among a predetermined set of alternatives, or among all possible distributions, or some specified set of distributions? In addition, how adequate is the fit as specified by the formal goodness of fit test cited?

In analyzing the failure behavior of the breakwater, the authors do a thorough job and provide some interesting insights. Here the only criticism is the selection of 500 as the suitable number of simulation replications. On what is this number based?

The development of the maintenance model also leaves some questions. There are well established procedures for maintenance optimization [see for example Sherif and Smith (1983)], why has the discount approach been undertaken. In addition, the development of the formula for K(ΔT) is not clear. The rest of the analysis is standard, though it may be improved by considering the affect of the parameter uncertainty on the uncertainty of the cost at the optimal solution. An overview of such an approach is given in Mazzuchi and Soyer (1995).

REFERENCES

[1] Lindley, D. V., (1983), "Reconsiliation of probability distributions", Operations Research, 31, 866-880.

[2] Mazzuchi, T. A. and Soyer, R. (1995), "Adaptive Bayesian replacement strategies", to appear in Proceedings of the Fifth Valencia International Meeting on Bayesian Statistics.

[3] Sherif, Y. S. and Smith, M. L., (1981), "Optimal maintenance models for systems subject to failure–a review", Nav. Res. Log. Quart., 28, 47-74.

[4] Vrijling, J. K. and van Gelder, P. H. A. J. M.,(1995) "Probabilistic design of berm breakwaters".

21

REPLY TO THE REVIEWS

Pieter van Gelder, and Han Vrijling

Department of Civil Engineering,
Delft University of Technology, Delft, The Netherlands

First of all, the authors want to thank the reviewers for their interesting discussions. We would like to give the following reply.

Dekker gives a good summary of the main points of interest from a problem owner perspective. He asks why figure 4 shows some kind of linear behaviour in the longshore transport over the breakwater's lifetime during monsoons. This linear effect is caused by the fact that in the simulations the realisations of B and D_{50} of the initial nourishment are kept constant during the breakwater's lifetime, whereas the realisations for the monsoon - and hurricane wave height are drawn every period of 3 and 24 hours respectively. If the initial nourishment is simulated with very low realisations of B and D_{50}, the breakwater will have a lot of erosion even under small storms (note that the monsoon wave heights have a relatively small variation coefficient in comparison with the hurricane wave heights).

We agree that other maintenance policies such as suggested in Dekker's review are worth considering. The periodic repair policy is chosen by the authors as a first approach to determine the optimal repair interval. It is used frequently in hydraulic engineering applications and is relatively easy to calculate.

Dekker points in his review to the problem of length scaling. We have only considered the upwind part of the berm breakwater consisting of R_0 stones and a probability of failure has been calculated for this section. Length scaling problems also appear in failure analysis of dikes; for a good overview we may refer to Calle (1988). The difficulty with length scaling for berm breakwaters is that subsequent sections have a very strong interaction via the longshore transport. Erosion in section i may result in accreation in section $i+1$, etc. To

R. Cooke et al. (eds.), Engineering Probabilistic Design and Maintenance for Flood Protection, 209–213.

model this correctly, the trajectory of each stone should be followed, taking into account their diameter, mass, the wave impact and many other variables. An averaging out of stone properties along the breakwater seems likely however.

In his review Van der Toorn gives a list of comments from engineering perspective. He starts with the problem of length scaling, for which we can refer to the above answer. He continues with comments on the level II calculation. He draws however an errorous conclusion from the numbers in table 1. The fact that the α_i^2 factors are small for the upper integration-bounds does not mean that the wave height has no influence on the outcome of Z. It does mean that the influence of these integration bounds on the total uncertainty in Z is of neglectable amount. Consider the reliability function $Z = R - S$ where Z is a function of n random variables X_1,X_2,...,X_n and R the resistance and S the sollicitation. We want to examine the probability of failure $Z = Z(X_1, X_2, ..., X_n) = 0$. In a level II analysis we are linearizing the Z function in a point X^* with the highest likelihood on the line $Z = 0$:

$$Z = Z(X_1^*, X_2^*, ..., X_n^*) + \Sigma \partial Z/\partial X_i(X_i - X_i^*) = 0 \tag{21.1}$$

In this point we can calculate its mean and standard deviation as follows:

$$\mu_Z = \Sigma \partial Z/\partial X_i(\mu_{X_i} - X_i^*) \tag{21.2}$$

$$\begin{aligned}\sigma_Z &= \sqrt{varZ} = \sqrt{\Sigma var(\partial Z/\partial X_i(X_i - X_i^*))} = \\ &\sqrt{\Sigma(\partial Z/\partial X_i)^2 varX_i} = \Sigma \alpha_i \partial Z/\partial X_i \sigma_{X_i}\end{aligned} \tag{21.3}$$

So α_i give us an indication of the relevance of stochast X_i on the total uncertainty in the probability of failure.

Van der Toorn continues with a suggestion that there is in fact no need for performing a level III analysis if from the level II analysis it follows that all the uncertainty in Z comes from the uncertainty in the parameter B. However, the level III analysis provides us with the exact probability of failure given all the parameter uncertainties, whereas the level II analysis is mainly used to examine the relevance of each parameter on the total uncertainty of the probability of failure.

Regarding the maintenance model, Van der Toorn tries to simplify the calculations for costs of repair and costs of failure but fails to take the discount factor into account. The suggestion to start simple with $r = 0$ might be attractive although the results from such a calculation can't be used in practice where you have to design structures of a life time over 50 years and when a capital of \$1000 corresponds to an initial investment of \$140 (at 4% interest). The use of discount factors is also advocated by Van Noortwijk (1996). A better explanation of the derivation of the formula for the total costs (15.8) is however worthwhile. The total costs equal the sum of "repair costs" and "risk costs". Every cost type has to be discounted. Because the repairs take place on the beginning of year $j = T$ ($j = 1..DIV(50, T)$), the first sum in formula (15.8) can be explained. After every repair the berm breakwater is in its original condition and its probability of failure is again given by pi in the ith year after repair ($i = 1..DIV(50, T)$). This explains the second sum in (15.8). The third sum was already explained in the main paper.

In his last three comments Van der Toorn ends with interesting suggestions for other maintenance strategies and for the optimization of the amount of initial nourishment.

In the third and final review a good examination from the mathematical perspective is given by Mazzuchi. He starts his review with a summary of all the assumptions that were made in the analysis and asks where they have their origin. Most of the assumptions are taken from Vrijling (1991) where a reference is made to empirical studies for the assumptions 4-8 and 10-14. The formula of assumption 9 is given in the case how to deal when the monsoon wave height might not be assumed constant in a 3 hours period. Assumption 16 is motivated by the limited depth in front of the breakwater.

We agree that the integration in formula (15.2) should be taken over the full range of the variables H and s (wave height and steepness). The reason for limiting the range between 0 and the maximum value of these variables in the design period of the berm breakwater was out of numerical convenience. Integrating further than this maximum value didn't have much effect on the expected longshore transport, other than increasing calculation time. The engineering judgment behind the bounded integration was to integrate the transport formula only over those wave heights and steepness that can occur in 50 years. Applying the transport formula for higher wave heights and steepness than those which can occur once every 50 years would be "dangerous", whereas this formula is not validated for extreme high wave height and steepness.

Mazzuchi continues with an analysis of the transport relation (15.1) and asks why the third parameter C isn't also considered as stochastic. The stochastic assumption for the parameters A and B was already good enough to express the uncertainty in the transport relation. The limited data set did not warrant the use of an extra random variable that would only mean extra calculation time. In the first analysis 2 pairs of (A,B) values where examined: $(4.8 \times 10^{-5}, 100)$ and $(7.2 \times 10^{-5}, 80)$. They were taken by engineering judgement on basis of model studies. With these fixed values of A and B, the integration in formula (15.2) is performed. With the choice of the second pair of A,B values, we refer to a safe or conservative transport relation because the designer opts for a heavier construction "just in case".

Regarding the fitting procedure for the data of S_{life}. From a set of possible distribution functions, its parameters have been estimated from the data of S_{life} with a least squares method. The fit with the lognormal distribution gave the lowest sum of squared residuals.

Mazzuchi concentrates on the level II analysis and would particularly like an explanation of the level II table. For every random variable X_i, its distribution type including its parameters is given. For the Gumbel and Frechet distributions also the μ and σ values are given for the approximating normal distribution in the design point X_i^* (see also reply on Van der Toorn's review). Finally the α_i^2 values are given for each stochast. A good reference on level II analysis is given by Thoft-Christensen and Baker (1982).

The authors welcome Mazzuchi's suggestions to apply a loss function setup for Z and to use Lindley's argument for the assumptions of normal distributions for unknown constants. Also the suggestion to improve the maintenance optimization by considering a Bayesian strategy is very much appreciated. In this regard we would like to point the follow-up paper of the berm breakwater analysis by Van Noortwijk and Van Gelder, where a Bayesian analysis is performed while taking into account the uncertainty of (i) the limiting average rate of initial rock displacement and (ii) given initial rock displacement has occured, the limiting average rate of longshore rock transport.

REFERENCES

[1] Van Noortwijk, J.M., (1996), "Optimal Maintenance Decisions for Hydraulic Structures under Isotropic Deterioration", PhD-thesis, Delft University of Technology.

[2] Van Noortwijk, J.M. and Van Gelder, P.H.A.J.M., (1995), "Optimal Maintenance Decisions for Berm Breakwaters", Technical Report 96-28, Delft University of Technology.

[3] Vrijling, J.K., Smit, E.S.P., and De Swart, P.F., (1991), "Berm breakwater design - the longshore transport case: a probabilistic approach", *Coastal Structures and Breakwaters*, London.

[4] Calle, E.O.F., (1988), "Een praktische benadering van de faalkans van een seriesysteem met identieke elementen en constante onderlinge correlatie", Grondmechanica Delft Rapport CO-294460/18, i.o.v. Rijkswaterstaat DWW.

[5] Thoft-Christensen, P. and Baker, M.J., (1982), *"Structural Reliability and its applications"*, Springer, Berlin.

REFERENCES

MIX
Papier aus verantwortungsvollen Quellen
Paper from responsible sources
FSC® C105338

If you have any concerns about our products, you can contact us on
ProductSafety@springernature.com

In case Publisher is established outside the EU, the EU authorized representative is:
Springer Nature Customer Service Center GmbH
Europaplatz 3, 69115 Heidelberg, Germany

Printed by Libri Plureos GmbH
in Hamburg, Germany